Extreme Climate Adaptation Strategy

Smart Water Grid in the Yangtze River Delta

East China Normal University Scientific Reports

ISSN: 2382-5715

This book series reports valuable research results and progress in scientific and related areas. Mainly contributed by the distinguished professors of the East China Normal University, it will cover a number of research areas in pure mathematics, financial mathematics, applied physics, computer science, environmental science, geography, estuarine and coastal science, education information technology, etc.

Published

Vol. 19 *Extreme Climate Adaptation Strategy: Smart Water Grid in the Yangtze River Delta*
by Heqin Cheng, Xi Chen and Kai Tan
(East China Normal University, China)

Vol. 18 *Topological Structure of Road Networks: Models and Computation Methods*
by Hong Zhang (East China Normal University, China)

Vol. 17 *Geography of Technology Transfer in China: A Glocal Network Approach*
by Chengliang Liu (East China Normal University, China)

More information on this series can also be found at https://www.worldscientific.com/series/ecnusr

(Continued at end of book)

East China Normal University Scientific Reports | Vol. 19

Extreme Climate Adaptation Strategy

Smart Water Grid in the Yangtze River Delta

Cheng Heqin
Chen Xi
Tan Kai

East China Normal University, China

World Scientific

NEW JERSEY • LONDON • SINGAPORE • BEIJING • SHANGHAI • HONG KONG • TAIPEI • CHENNAI • TOKYO

Published by

World Scientific Publishing Co. Pte. Ltd.
5 Toh Tuck Link, Singapore 596224
USA office: 27 Warren Street, Suite 401-402, Hackensack, NJ 07601
UK office: 57 Shelton Street, Covent Garden, London WC2H 9HE

Library of Congress Control Number: 2026001718

British Library Cataloguing-in-Publication Data
A catalogue record for this book is available from the British Library.

East China Normal University Scientific Reports — Vol. 19
EXTREME CLIMATE ADAPTATION STRATEGY
Smart Water Grid in the Yangtze River Delta

ISBN 9789819827206 (hardcover)
ISBN 9789819827213 (ebook for institutions)
ISBN 9789819827220 (ebook for individuals)

For any available supplementary material, please visit
https://www.worldscientific.com/worldscibooks/10.1142/14701#t=suppl

Desk Editors: Soundararajan Raghuraman/Amanda Yun

Typeset by Stallion Press
Email: enquiries@stallionpress.com

Preface

The Yangtze River Delta holds a crucial position in the economic landscape of China and even the world. It suffers from floods, droughts, and sea level rise, causing environmental changes with significant risks to potable water, flood prevention, and tide control, impacting port and shipping activities, all of which are closely related to the water network. Recently, extreme weather and climate events have increased in frequency and intensity, often occurring in clusters and concurrently, posing a serious threat to regional water resources security due to more severe saltwater intrusion, water supply crises, degradation of the water environment and the water ecosystem, and disruptions to inland river shipping. Existing resilience and resistance systems and the current level of forecasting are inadequate to handle extreme climate events. This book focuses on risk assessments and predictions of extreme climate changes from the perspective of big data and artificial intelligence technology.

The development of a smart water grid, a "digital water network," can meet application needs and achieve the goals of real-time monitoring, rapid transmission, accurate forecasting, optimized decision-making, precise allocation, and efficient management. It provides support for precise allocation, optimized decision-making, and efficient management of water resources during floods and droughts. The construction of a smart and high-quality water grid in the Yangtze River Delta, incorporating independent intellectual property rights, will ensure profound beneficial effects on the strategic countermeasures for the sustainable development of water resources urgently needed by world-class strategic emerging industrial clusters in the Yangtze River Delta and promote the national development strategy of regional integration in the Yangtze River Delta. Here is a brief introduction to the topics covered in each chapter.

Chapter 1 clarifies the definition and challenges of water networks for water supply, flood control, navigation, and ecological functions in response to frequent extreme climate events in recent years in the Yangtze River Delta. The importance and urgency of developing smart water networks in the region are also emphasized for policy makers.

Chapter 2 introduces the ongoing research that has been conducted, focusing on the significant basic scientific issues related to the normalization of extreme climate and the magnitude of relative sea-level rise in estuarine deltas, as well as their impacts on intelligent monitoring, identification, and early warning of disaster risks concerning urban flooding and tide control, water transportation and supply, and ecological security, based on the original achievements obtained in recent years.

Chapter 3 addresses the current status and problems of the water network in the Yangtze River Delta; this discussion elaborates on the necessity, guiding directions, benefits, governance challenges, and long-term mechanisms for responding to the normalization of extreme weather through the construction of a smart and high-quality water grid in the Yangtze River Delta.

Chapter 4 presents the current challenges and key issues of the water network management platform in the Yangtze River Delta. A plan is proposed for building an intelligent and high-quality water network management platform in the Yangtze River Delta, considering the context of normalized extreme climate. It is composed of the construction of an “air–space–ground–sea” Internet of Things (IoT) network and a continuous water quantity and quality monitoring system with sensor placement.

Chapter 5 elaborates on the foundational and prerequisite elements for the effective operation of a smart, high-quality water network, which includes the intelligent data collection regarding physical and chemical elements, data processing, and data analysis of basic information using products locally developed in China with domestic intellectual rights, as well as capacity building for regional autonomy in the sustainable utilization of water resources in the Yangtze River Delta.

Chapter 6 proposes a new method to estimate terrain elevations at 10 m resolution by integrating a random forest machine learning model and remotely sensed data from ICESat-2, Sentinel-1, and Sentinel-2 satellites. This enables us to accurately map terrain elevations at a large scale and fine resolution, characterizing detailed surface height and geomorphic changes, critical for studying the internal motions and external forces of the Earth. This is the fundamental geomorphological element of the smart water network. The core principle is to construct a random forest model that characterizes the complex relationships between the ICESat-2 ATL03 terrain elevations and their corresponding land cover-related polarization characteristics and spectral variables derived from Sentinel-1 and Sentinel-2, respectively. It ensures satisfactory robustness performance of the proposed method under different land covers and relatively good consistency between the predicted elevations and the actual terrain of the entire island.

Chapter 7 elucidates an effective and intelligent way to achieve more accurate classification and regression of dissolved oxygen distribution trends in the Taihu Lake Basin using a multi-scale temporal fusion transformer based on physical and chemical prior knowledge. This is because dissolved oxygen (DO) is a key parameter for monitoring water quality, as it provides important guidance for aquaculture and environmental warnings. This study proposes a model for predicting and classifying DO concentration in the Taihu Lake Basin, based on a prior knowledge-guided multi-scale temporal fusion transformer (PKMSTFT) model, aiming to classify and predict DO concentration at 15 hydrological stations within the basin. Furthermore, Shapley additive explanations (SHAP) are used to analyze the impact of internal parameters on DO. To further validate the effectiveness and competitiveness of the PKMSTFT model, we compared it with nine other commonly used machine learning algorithms, including convolutional neural networks (CNNs) and Autoformer. The results show that the overall performance of the PKMSTFT model is superior to the other models in nine datasets. This method can provide a reference and theoretical basis for the scientific management and monitoring of water resources.

Chapter 8 discusses the comprehensive natural, economic, and social risk assessment and countermeasures carried out in response to the severe water supply challenges faced by the Shanghai mega-city and other estuarine delta cities, particularly in the face of escalating extreme climate events. This research employs system dynamics and dynamic adaptive policy pathways to assess Shanghai's water supply security risks and responsive strategies from 2022 to 2050, amid extreme climate conditions. Using data from 2000 to 2021, we constructed a system dynamics model to predict future water demand under various development scenarios. Focusing on the unusual 2022 drought in the Yangtze River Basin, we simulated 15 scenarios, including economic, population, water efficiency, and reservoir levels, to identify the extent and timing of potential water supply risks and then proposed pertinent dynamic adaptive strategies to address them. Short-to-medium-term recommendations include optimizing both local and transit water resources, strengthening emergency water reserves, enhancing water use efficiency, and maintaining stable reservoir water levels. For the long term, expanding water storage infrastructure and promoting integrated water resource management within the Yangtze River Delta are key to establishing a resilient and diversified water supply system, effectively mitigating future water security risks.

Chapter 9 describes the main suggestions for future research regarding the construction of smart, high-quality water network infrastructure in the Yangtze River Delta to address the challenges posed by extreme climate change and emergencies. It involves establishing a comprehensive atlas for water networks early on to adapt to new challenges in National Water Grid Construction and strengthening the micro-storage and micro-circulation capabilities of Shanghai's modern water grid.

All in all, this book can serve as a valuable reference for researchers and professionals in the fields of geography, oceanography, meteorology, water resources engineering, and disaster risk reduction. It is also suitable as a reference text for graduate students and advanced undergraduate students studying relevant disciplines at universities. The academic value of this book resides in its strategic vision for developing a smart and high-quality water grid in the Yangtze Delta that is resilient to extreme climate change. We leverage national and Shanghai-based strengths in multidisciplinary frontier science and technology fields such as aerospace, information technology, geography, hydrology, and oceanography. This includes integrated monitoring of sea level variations, geological features, geomorphology, water quality, and ecological aspects of the water network in the Yangtze Delta and surrounding marine regions, as well as the advancement of air–space–ground–sea integrated network transmission systems and associated data processing equipment. An integrated air–space–ground–sea grid solution must be established for the Yangtze Delta and its adjacent marine areas.

Perspectives are outlined regarding the development of a decision-analysis system for the smart, high-quality water grid in the Yangtze Delta. These perspectives encompass advancements in prediction technologies for the spatio-temporal distribution of water resources and potable water supply within the Yangtze Delta, utilizing sophisticated spatio-temporal sequence prediction models. They also cover online assessments and strategic response methodologies for the effects of extreme weather events, meteorological occurrences, and emergencies on the smart water grid. Additionally, they propose data products and models for real-time evaluation and early warnings of safety risks in the high-quality water grid of the Yangtze River Delta. This includes leveraging AIoT for continuous monitoring, integrating air–space–ground–sea multi-source heterogeneous remote sensing data, online sensor readings, and generating high-precision spatio-temporal datasets pertinent to the water grid at both macro and micro scales. These insights suggest combining water environment and water ecological disaster models—such as those for algal blooms and saltwater intrusion—with the development of an offline water resources scheduling rule library. This library, grounded in deep learning techniques, would archive resource allocation protocols and strategies tailored to diverse scenarios. Resource scheduling can then be initiated in response to factors such as population density shifts or disaster responses.

About the Authors

Cheng Heqin, PhD, is a researcher professor and doctoral supervisor at the State Key Laboratory of Estuarine and Coastal Research, East China Normal University (ECNU). She has extensive experience in conducting comprehensive analyses that integrate field measurements from space-based, shore-based, manned vessel-based, and unmanned boat-based multimodal sensors with historical data, numerical models, analytical models, and deep learning. Her research focuses on the impacts of the fundamental physical processes of continuous hydrodynamic systems stretching from large river basins to estuaries and oceans, under the combined effects of climate warming, sea-level rise, and intense human activities, on major infrastructure such as flood control and transportation along rivers and coasts, as well as on freshwater and fishery resources. Her primary research areas include dynamic sedimentary geomorphology, underwater engineering geomorphology, and integrated coastal zone management.

Chen Xi, PhD, is a professor and doctoral supervisor at the Institute of Software Engineering, ECNU, China. He received his PhD from Shanghai Jiao Tong University, China, and completed his postdoctoral research at Purdue University, USA. He was formerly a lecturer at Harbin Institute of Technology. China. His primary research focuses on artificial intelligence and data analysis applications. He has contributed (as a key participant) to China's national projects under the 973 and 863 Programs, and has led projects funded by the National Natural Science Foundation of China. He has published over 40 papers (in English) in prominent international journals and conferences, holds nearly 20 granted invention patents, and has authored a monograph supported by China's National Publication Fund.

Tan Kai, PhD, is an associate research professor at the State Key Laboratory of Estuarine and Coastal Research, ECNU, China. He received his BE degree in surveying engineering from Wuhan University, China, and PhD degree in surveying and mapping from Tongji University, China. He was previously a research assistant at the Hong Kong Polytechnic University, Hong Kong, China. His major research interests include multi-source remote sensing technologies and their applications in the coastal zones. Kai Tan has published over 50 academic papers on topics such as delta topography, coastal vegetation, and estuary engineering. He has conducted over 20 international and national projects as principal investigator (PI) and co-PI. He was selected for the "Young Science and Technology Talents Sailing Plan" of the Shanghai Talent Project and won the "Guangdong Provincial Science and Technology Progress Award", and the "Guangdong Engineering Survey and Design Industry Association Science and Technology Award". As a member of the Education Working Committee of the Chinese Society of Surveying and Mapping and the LiDAR Professional Committee of the China National Committee of the International Digital Earth Society, he is dedicated to promoting interdisciplinary research and collaboration.

Acknowledgments

This research was supported by Shanghai's "Science and Technology Innovation Action Plan" for the Year 2023: Targeted Commissioned Projects in the Field of Soft Science Research 23692123900, the National Natural Science Foundation of China (Nos. 42271009 and 51761135023), and Fundamental Research Funds for the Central Universities, Shanghai Natural Science Foundation (No. 22ZR1420900).

First of all, we would like to express our sincere thanks to Professor Qian Xuhong, president of East China Normal University and an academician of the Chinese Academy of Engineering. Your rigorous supervision and full support throughout the project's initiation, execution, and final acceptance stages have made its successful completion possible. We are also deeply grateful to Professor Zhang Jing, academician of the Chinese Academy of Sciences, for his strong support. We also extend our sincere thanks to Professor Ye Shuhua, academician of the Chinese Academy of Sciences, for her long-term guidance and support.

We would also like to express our gratitude to Professor Li Jiren from the China Institute of Water Resources and Hydropower Research; Professor Li Deren, academician of the Chinese Academy of Sciences and the Chinese Academy of Engineering and a professor at Wuhan University; Professor Pan Delu, academician of the Chinese Academy of Engineering from the Second Institute of Oceanography, Ministry of Natural Resources; Professor Zhou Chenghu, academician of the Chinese Academy of Sciences at the Institute of Geographic Sciences and Natural Resources Research,

Chinese Academy of Sciences; Professor Wan Haibin, former inspector specialist of the Flood and Drought Disaster Prevention Department of the Ministry of Water Resources; Professor Chen Junping and Professor Yang Zhigen, and Associate Professor Li Jin from the Shanghai Astronomical Observatory, Chinese Academy of Sciences; Engineer Qiu Luyan from the Shanghai Academic Activity Center of the Chinese Academy of Engineering; Professor Zhang Xianlin, former deputy chief engineer of the Shanghai Municipal Planning and Natural Resources Bureau; Professor Zhou Fengnian, former director of the Changjiang Estuary Hydrology and Water Resources Survey Bureau of the Changjiang Water Resources Commission Hydrology Bureau; Professor Lin Hejuan, former deputy director of the Hydrology Bureau of the Taihu Lake Authority; Professor Xu Guiquan, former director of the Shanghai Water (Ocean) Planning and Design Institute; Professor Zhou Quanping, Nanjing Geological Survey Center, China Geological Survey, China; Professor Han Zhen, Shanghai Ocean University; Professors Zhu Jianrong, Zeng Gang, Zhu Liangfeng, Xu Jing, He Liang, Li Chao, Liu Min, Yu Boliang, Gao Hongkai, and Yao Yuanzhi, and Teacher Jia Ye from East China Normal University.

Contents

Chapter 1

Introduction

The water network is the physical space on which water flow depends, and it stands alongside the transportation, energy, and communication networks as one of the four major infrastructure networks of civil engineering in modern society. It is the only irreplaceable network and the most dynamic controlling factor in ecosystems. As human society has developed to the present, the water network has evolved into a complex integration of natural rivers, lakes, and seas, incorporating various water conservation facilities. Based on natural water systems, with water diversion, regulation, and drainage projects serving as channels and node control projects serving as hubs, it is a water distribution network system that integrates flood and drought prevention and control, water resources allocation, water ecology protection, and water transportation. Composed of natural river and lake systems and multiple engineering components, it is a water resources network, a water ecology network, a water energy network, and a water transportation network. It is a network system with composite functions and complex structures, involving challenging scheduling requirements, and it exhibits significant spatiotemporal scale effects.

During a symposium on promoting the high-quality development of the South-to-North Water Diversion Project on May 4, 2021, General Secretary Xi Jinping clearly stated the need for a "complete, safe, reliable, intensive, efficient, green, intelligent, smooth-circulating, and orderly-regulated" national water network by accelerating the construction of infrastructure. This is a water-related activity that spans watersheds and regions, representing a complex and large-scale system. To build the national water network, it is essential to fully leverage the new generation of information technology to digitally empower all aspects of water network planning, design, construction, and operation. This will enhance the digitization, networking, and intelligence of the national water network, providing robust water security safeguards for the comprehensive development of a modern nation. Implementing major national water network projects is one of the six implementation pathways to achieve water security in China.

The national water network is a crucial infrastructure for comprehensively addressing issues related to water resources, water ecology, water environments, and water disasters, as well as for comprehensively enhancing the national capacity for water security guarantee. On the one hand, regional water networks connect with the national backbone water network; on the other hand, they enhance regional water security and support major national development strategies, such as the development of the Yangtze River Economic Belt and the integrated development of the Yangtze River Delta.

The Yangtze River Delta region occupies an important position in China's and even the world's economic landscape. It covers three provinces and one municipality—Shanghai, Jiangsu, Zhejiang, and Anhui—supporting a permanent resident population of 237 million people. With less than 4% of the country's land area, it has consistently accounted for about one-quarter of the country's economic output over many years (Figure 1.1).

However, the region lacks sufficient water resources to support high-quality economic and social development. The per capita water resource volume is only 940 cubic meters, far below the world average. The quality of raw water for urban and rural water supplies is low, and the supplies' ability to resist risks is weak. During the severe saltwater intrusion period in 2022, Shanghai's Qingcaosha Reservoir was unable to draw water for 98 consecutive days, far exceeding the design standard of 68 days, and some areas experienced prolonged and continuous droughts. Shanghai's Chenhang Reservoir experienced 11 saltwater intrusions between September 2006 and March 2007. Some areas have poor water network connectivity; flood disasters in the Taihu Lake basin and Huai River basin have consistently posed a threat; there is considerable pressure on the water environment and ecological protection; and shipping capacity is insufficient. The demand for regional water network construction is evident in

Figure 1.1. Sketch map of Yangtze Delta cities.

this region. The construction of a high-quality water network in the Yangtze River Delta is both necessary and urgent. At the same time, with favorable conditions, its feasibility is clearly advantageous.

In recent years, global warming has become evident. The global average temperature has increased at an average rate of 0.15°F per decade since 1901, reaching the highest temperature on meteorological record in 2023 and projected to increase by 1.5–2.8°C by the end of this century. The surface temperature of the world's oceans has gradually risen since the beginning of the 20th century, increasing by 0.13°F per decade from 1901 to 2015. From 1880 to 2013, sea levels rose at a rate of 1.52 mm per year. Beginning in 2014, this rate increased to 2.79–3.81 mm per year, almost doubling the previous rate. Meanwhile, extreme weather and climate events are exhibiting new trends of increasing frequency, intensity, clustering, and concurrency. Some historically rare and extreme events are becoming more normalized, posing a severe threat to regional water security. Global water issues, such as flood losses, water supply crises, water environment and ecosystem damage, and hindered inland navigation, have become more severe. For example, in 2020, the Yangtze River basin experienced a major flood event, second only to those in 1954 and 1998 since the founding of the People's Republic of China. Severe droughts occurred in the Yangtze River basin in 1961, 1978, and 2022, and the possibility of extreme drought events similar to those that occurred in 1627–1643, 1874–1879, and 1922–1932 cannot be completely ruled out.

Given that existing flood control and drought relief systems, as well as ecological protection systems, are difficult to manage in extreme climate events, and current forecasting capabilities cannot achieve ultra-long-term forecasts, modern urban life and industries are ill-equipped to withstand extreme droughts and floods. Therefore, we need to adopt a limit-thinking approach to enhance the adaptability of the water resource system in the Yangtze River Delta region to the normalization of extreme weather events. The smart water grid is a "digital water network" that integrates the "spatial data water network, process water network, and topological water network." It can be built and iteratively evolved with the support of new-generation information technologies such as big data, cloud computing, the Internet of Things (IoT), and AI. It provides different manifestations of the water network to meet the application needs of the actual business layer, ultimately achieving the goals of "real-time monitoring, rapid transmission, accurate forecasting, optimized decision-making, precise allocation, and efficient management." It supports precise allocation, optimized decision-making, and efficient management of water resources during floods and droughts. Accelerating the construction of a smart, high-quality water grid in the Yangtze River Delta, with independent intellectual property rights, ensures the implementation of engineering countermeasures and strategies for sustainable water resource development that are urgently needed by world-class strategic emerging industrial clusters in the region, thereby promoting the national development strategy of regional integration in the Yangtze River Delta.

To this end, the Shanghai Municipal Science and Technology Commission launched the soft science consultation project "Strategic Research on the Construction of a Smart and High-Quality Water Grid in the Yangtze River Delta under the Normalization of Extreme Climate" in June 2023. The project, involving East China Normal University, the Shanghai Astronomical Observatory of the Chinese Academy of Sciences, and the Shanghai Academician Center of the Chinese Academy of Engineering, conducted in-depth research and interviews on topics such as "Online Monitoring and Prediction of Relative Sea Level Changes in the Yangtze River Delta," "Online Monitoring, Prediction, and Allocation Plan for Water Resources in the Yangtze River Delta and Water Scarcity in Shanghai under the Normalization of Extreme Climate," and "Construction of a Management Platform for a Smart and High-Quality Water Grid in the Yangtze River Delta under the Normalization of Extreme Climate," according to the project plan. After thoroughly investigating and summarizing the flood control and water supply situations of major international megacities with river estuaries such as New York, Greater Vancouver, London, Paris, Sydney, and Tokyo, as well as the roles and issues of existing water conservancy projects such as the Foziling Reservoir in Huaibei and the Qingcaosha Reservoir in Shanghai, the control gates of the Taipu River on the Huangpu River in Shanghai, the north and south diversion rivers on Chongming Island, the sluice gates and dams along the Shaying River in Huaibei region, Anhui Province, the water diversion projects from the Yangtze River to the Huaihe River and Taihu Lake, and the Pishihang Irrigation Area, in addition to the needs of flood control and drought relief for water network construction, the project members together convened the "Strategic Discussion and Academician Expert Consultation Meeting on the Construction of a Smart and High-Quality Water Grid in the Yangtze River Delta" on March 5, 2024. Two decision-making consultation opinions from academicians and experts were put forward: "Suggestions for Promptly Developing and Constructing the Infrastructure, Risk Threshold Perception, and Analysis System for the Smart and High-Quality Water Grid in the Yangtze River Delta" and "Early Initiation of Constructing a Comprehensive Atlas for Water Grid to Address New Challenges in National Water Network Construction and Strengthening the Micro-Storage and Micro-Circulation Capabilities of Shanghai's Modern

Water Network." These opinions specifically cover the following three important aspects:

1. To develop mechanisms as early as possible to clarify the sharing of responsibilities among the main entities of the Yangtze River Delta water network, enhance regional policy coordination, strengthen incentives for innovation breakthroughs, and expand global cooperation networks.
2. To promptly initiate research and development of the risk threshold perception and analysis system for the infrastructure of the smart, high-quality water grid in the Yangtze River Delta.
   (i) To research and develop monitoring and data transmission technologies and equipment for the physical elements of the integrated smart water grid encompassing air, sky, land, and sea.
   (ii) To fully leverage the advantages of multidisciplinary research, industry-academia-research collaboration, and cutting-edge scientific and technological fields in aerospace, information, geography, water affairs, marine sciences, and related fields in China, especially in Shanghai. This includes the integrated monitoring of sea level changes, flood disasters, geological landforms, water quality, and ecological elements within the Yangtze River Delta water network and adjacent sea areas, as well as research and development of network data transmission and equipment. Establish an integrated network solution for air, sky, ground, and sea within the Yangtze River Delta and adjacent sea areas, incorporating a Chinese space reference antenna and a prototype terminal for low-orbit satellites, integrating communication, sensing, and remote sensing capabilities. Develop standards for the transmission of sensing data in the Yangtze River Delta and adjacent sea areas, and implement them as soon as possible to support integrated prevention of water and drought disasters in both land and sea areas.
   (iii) To research and develop a decision-making analysis system for the smart, high-quality water grid in the Yangtze River Delta.

   Several important measures need to be implemented as soon as possible: The first is to develop predictive technologies for the spatiotemporal distribution of water resources and water supply volumes in the Yangtze River Delta, based on large-scale spatiotemporal sequence prediction models. Second, develop online assessment and strategic response technologies for the impact of extreme climatic and meteorological events and emergencies on the smart, high-quality water network. Third is to research and develop data products and models for online assessment and early warning of security risks in the high-quality water network of the Yangtze River Delta. These should include AIoT-based online monitoring and collaborative multi-source heterogeneous remote sensing of big data from the air, sky, ground, and sea, as well as online sensor monitoring data and high-precision spatiotemporal data products for the Yangtze River Delta water network at both macro and local scales. Fourth involves combining these elements with water environment and ecological disaster models, such as algal blooms and salt-water intrusions, and an offline water resource scheduling rule library based on deep learning to store resource scheduling rules and strategies for various scenarios. Resource scheduling can be triggered through the identification of events such as population density and disaster emergencies.

(iv) To expand the application services of the smart, high-quality water grid in the Yangtze River Delta.

This includes several important steps. The first step is to analyze and predict medium- and long-term water demand scenarios in the Yangtze River Delta and to establish an intelligent water dispatch system. The second step is to develop a perception network for waterway conditions, relying on the smart and high-quality water network in the Yangtze River Delta. Third is to build a platform for sensing, comprehensive monitoring, early warning, and forecasting of security risks within the smart and high-quality water network in the Yangtze River Delta.

3. To urgently initiate planning and research on micro-storage and micro-circulation systems for the modern water network in Shanghai.

Starting from modern water science research informed by big Earth data, extensive knowledge, and large models, we emphasize the deep intersection and integration of water science with information science, geography, mathematics, and other disciplines. By exploring the internal relationships within the water network system of the estuarine delta metropolis (Shanghai), which features complex dominant factors, hierarchical systems, and intricate correlations, we aim to establish key theories and methodologies suitable for constructing water network atlases across multiple spatiotemporal scales, hierarchical structures, and multi-objective domains. We further aim to overcome fundamental scientific issues such as the representation of water network atlases in metropolitan areas, high-precision digital water network timelines, and evolutionary principles. Based on these foundations, we intend to construct a modern water network system in Shanghai that integrates "village-level capillary water networks with national artery water networks." This will enable the connection, integration, and fusion of village-level capillary water networks with their primary functions of micro-storage and micro-circulation, and national-level artery water networks. Key theoretical methodologies will be developed for the evolution of water network atlases during the rapid development of estuarine delta metropolitan areas, followed by validation tests in Shanghai. We hope to comprehensively advance the quality improvement project for Shanghai's water network, ultimately facilitating Shanghai's achievement of the three major goals of water resource security, freshwater supply security, and water ecological security.

To sum up, the Yangtze River Delta holds a crucial position in the economic landscape of China and even the world. It suffers from floods, droughts, and sea level rise, causing environmental changes with significant risks to potable water, flood prevention, and tide control, impacting port and shipping activities, all of which are closely related to the water network. Recently, weather and climate events have increased in frequency and intensity, occurring in clusters and concurrently, posing a serious threat to regional water resources security due to more severe saltwater intrusion, water supply crises, degradation of the water environment and ecosystem, and disruptions to inland river shipping. Existing resilience and resistance systems and the current level of forecasting are inadequate to handle extreme weather events. This book focuses on risk assessments and predictions of extreme climate change, from the perspective of big data and AI technology. The development of the smart water grid—a digital water network—can meet the application needs and achieve the goals of real-time monitoring, rapid transmission, accurate forecasting, optimized decision-making, precise allocation, and efficient management. It provides support for precise allocation, optimized decision-making, and efficient management of water resources during floods and droughts. The construction of a smart and high-quality water grid in the Yangtze River Delta, secured with independent intellectual property rights, will ensure a profound beneficial effect on strategic countermeasures for the sustainable development of water resources urgently needed by the world-class strategic emerging industrial clusters in the Yangtze River Delta and promote the national development strategy of regional integration in the Yangtze Delta.

Chapter 2

The Impact of Global Extreme Climate Change and Human Activities on the Water Network of the Yangtze River Delta

Water is a vital resource necessary for the survival and development of all life on Earth. As the global population continues to grow, demand for clean water resources is also increasing, intensifying competition for this scarce resource. Although water covers approximately 70% of the Earth's surface, only a small portion—about 3%—is freshwater, and much of it is not easily accessible or suitable for consumption (Ibrahim *et al.*, 2022). Therefore, the effective management of sustainable water resources has become a major concern (Foroumandi *et al.*, 2023).

The acceleration of global warming and sea level rise has become a consensus within the international community (China Meteorological Administration Climate Change Center, 2020; Ministry of Natural Resources Marine Forecast and Monitoring Division, 2020). The year 2023 was the hottest on record. How estuarine delta regions, with dense populations and economies, respond to these challenges is a major global frontier issue. Building on original results from recent years, this project continues to explore critical foundational scientific questions related to the normalization of extreme climate events and relative sea level rise in estuary delta regions and their impacts on urban flood defenses, water transportation and supply, and ecological safety through intelligent monitoring, identification, and early warning of disaster risks. The relevant research progress, both domestically and internationally, is as follows.

2.1 Overview of Water Resources in the Yangtze River Delta Region

According to the latest 2021 National and Jiangsu-Zhejiang-Shanghai-Anhui Water Resources Bulletin, the total water resources in the Yangtze River Delta region amounted to 278.265 billion m^3, accounting for 9.39% of the national total. Surface water resources totaled 260.943 billion m^3, accounting for 9.22% of the national surface water, while groundwater resources were 61.72 billion m^3, with a non-overlapping resource volume of 17.322 billion m^3, accounting for 13.05% of the national groundwater. Among the provinces, Zhejiang had the largest total water resources, with 134.473 billion m^3, and surface water resources of 132.333 billion m^3. Anhui possessed the richest groundwater resources, with a non-overlapping resource volume of 8.525 billion m^3 (Figure 2.1). The long-term average total water resource of the Yangtze River Delta is 208.845 billion m^3, and in 2021, the region experienced a 33.2% surplus compared to the long-term average.

By river basin, the total water resources of the Yangtze River Basin in 2021 amounted to 59.065 billion m^3 (excluding the Taihu Basin), with surface water resources of 56.782 billion m^3 and groundwater resources of 2.283 billion m^3. The Taihu Basin's total water resources were 27.586 billion m^3, with 25.487 billion m^3 of surface water and 2.099 billion m^3 of groundwater. The Huai River Basin possessed 62.525 billion m^3 of water resources, comprising 51.064 billion m^3 of surface water and 11.461 billion m^3 of groundwater. The Southeastern River Basin featured the most abundant water resources, totaling 129.088 billion m^3, with surface water at 127.609 billion m^3 and groundwater at 11.461 billion m^3. In summary, the Southeastern River Basin possessed the highest total water resources and surface water volume, while the Huai River Basin had the highest groundwater volume.

In 2021, the total water supply in the Yangtze River Delta reached 108.314 billion m^3, with surface water supplying 103.092 billion m^3

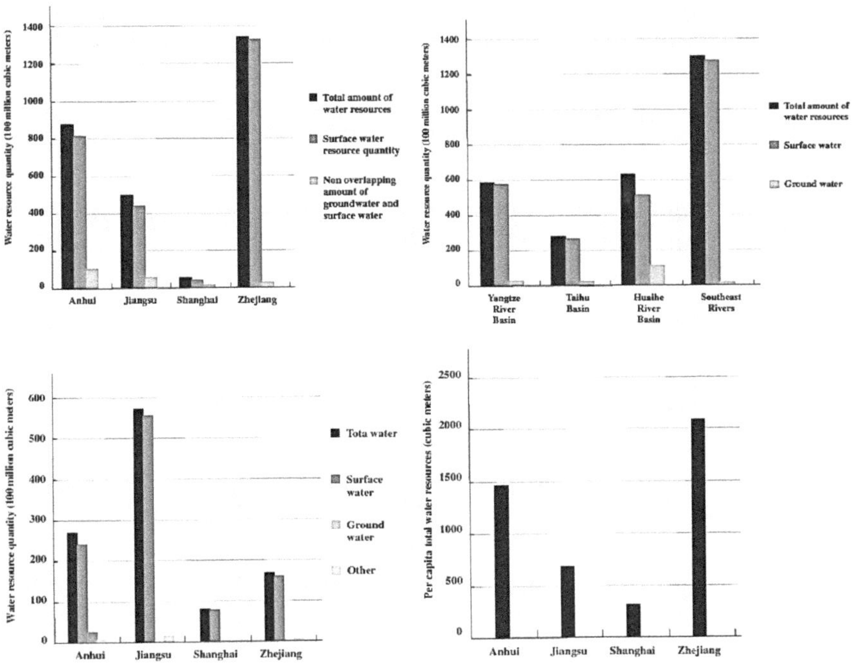

Figure 2.1. Water resources quantity in the Yangtze Delta.

and groundwater supplying 2.924 billion m^3. The overall water supply structure was predominantly surface water, though groundwater accounted for a higher proportion in some regions. In Shanghai, Zhejiang, and Jiangsu, surface water was the primary source, making up over 97% of the total water supply. In Anhui, surface water contributed 88.16% of the total, while groundwater supply reached 2.581 billion m^3, or 9.5%, with the northern Anhui region having a particularly high reliance on groundwater. In Suzhou, Huaibei, Bozhou, Fuyang, and Bengbu, groundwater made up 71.58%, 62.77%, 59.92%, 40.80%, and 15.73% of the water supply, respectively. In other cities, groundwater generally accounted for less than 2%. The proportion of groundwater usage has been decreasing year by year.

The Yangtze River Delta has abundant water resources overall, but there are significant differences between the northern and southern regions. The southern areas have larger quantities of high-quality water, while resource and water quality shortages persist in northern Jiangsu and northern Anhui. Due to the high population density, per capita water resources are relatively low, falling below the national average of 2,097.63 m^3per person. Among the Yangtze River Delta regions, Zhejiang has the highest per capita water resources at 2,056.17 m^3, while Shanghai has the lowest at just 311 m^3. As a result, due to the ongoing trend of extreme climate normalization, water resources in the Yangtze River Delta, especially in Shanghai, will face significant challenges in the future.

2.2 Overview of the Global Normalization of Extreme Weather Events

Under the influence of human activities, driven primarily by greenhouse gas emissions, atmospheric CO_2 concentrations increased by more than 50% compared to pre-industrial levels by 2023. The enhanced greenhouse effect has raised the global average near-surface temperature by approximately 1.45°C above pre-industrial levels, leading to more frequent and intense extreme weather events, such as heatwaves, heavy rainfall, floods, and droughts. According to the National Climate Center, since the beginning of the 21st century, China has experienced over 50 significant rainstorm and flood events and more than 20 major typhoons, severe droughts, and extreme heatwaves. This series of observational evidence highlights that China's weather and climate are shifting toward a new normal, characterized by frequent and intense extreme events, posing challenges to regional ecological security, water resource security, and the sustainable development of society and the economy.

As global warming continues to intensify, climate system models predict that extreme weather events will become even more severe and occur with greater frequency. For example, extreme rainfall and droughts, closely related to regional water resource security, are expected to increase significantly. Projections indicate that when the global average temperature rises by 2°C above pre-industrial levels, the frequency of what is currently considered a once-in-a-century heavy rainfall event will increase by nearly 50%, making it a once-in-50-years event. Although there is still significant uncertainty regarding future changes in extreme droughts, it is a well-established scientific understanding that the frequency, intensity, and duration of droughts will increase. In the Yangtze River Delta urban agglomeration, due to its unique geographical location, the likelihood of multiple extreme events occurring simultaneously or successively—such as heavy rainfall, floods, typhoons, and storm surges—will also rise, becoming a prominent feature of the region's new climate normal in the future.

Therefore, to build and maintain a smart water network system that adapts to the new normal of increasingly frequent and intense extreme events, it is essential to accurately predict changes in those events over the next 10, 30, and 50 years. This will ensure that the water network system can adapt to the extreme climate conditions of different future periods. Currently, medium- and long-term climate forecasts rely primarily on data from the Global Coupled Model Intercomparison Project, coordinated by the United Nations Intergovernmental Panel on Climate Change (IPCC). However, the low resolution of global models cannot meet the high-resolution extreme climate prediction requirements needed to ensure the safety of the smart water network in the Yangtze River Delta region. Considering the region's specific conditions, developing advanced artificial intelligence (AI) technologies to downscale data from the global coupled model intercomparison project and integrating AI with physical models to build a high-resolution extreme climate prediction system for the Yangtze River Delta is crucial. This approach will not only ensure the efficient operation of the smart water network system but also provide new pathways for high-precision extreme climate predictions in the AI era.

2.3 Extreme Drought and Flood Disaster Predictions for the Yangtze River Delta Region

Against the backdrop of global climate normalization, China frequently experiences floods and droughts, which are becoming increasingly harmful and causing greater losses. These disasters have become a prominent issue in flood prevention and disaster mitigation. Each year, droughts and floods lead to a loss of approximately 20 million tons of grain and over 200 billion yuan in economic damages, severely affecting water resource utilization and causing a series of social and environmental problems (Huang and Du, 2010; Qin, 2015). Therefore, accurate river runoff forecasting can provide crucial guidance for optimal water resource allocation, flood prevention, and disaster mitigation (Alizadeh *et al.*, 2017; Wu *et al.*, 2023; Swain and Patra, 2017). However, since runoff is significantly influenced by precipitation and human activities, it exhibits high variability and volatility, posing significant challenges to accurate hydrological predictions (Ha *et al.*, 2021; Herath *et al.*, 2021; Song *et al.*, 2012). Thus, the most urgent task is to develop a robust and precise hydrological model and apply it to practical use.

Runoff prediction methods can generally be divided into two categories: process-driven methods (Jaberzadeh *et al.*, 2022; So *et al.*, 2020; Wanders *et al.*, 2019) and data-driven methods (Kratzert *et al.*, 2019; Xie *et al.*, 2023; Yin *et al.*, 2021). Process-driven methods are based on a deep understanding of the physical mechanisms governing water behavior within the hydrological cycle. The primary goal of this approach is to describe these physical mechanisms in order to better understand and predict the movement of water through various components of the system, including precipitation, evaporation, infiltration, and surface and subsurface runoff. Although hydrological models offer high prediction accuracy and interpretability, simulating runoff using process-driven models involves multiple factors, such as climate, soil, and vegetation. These factors are complex and interrelated, making it challenging to estimate the impact of each on runoff. Furthermore, these factors are influenced by unpredictable elements, such as rainfall variability and temperature fluctuations (Liu *et al.*, 2022; Zhang *et al.*, 2021).

Compared to process-driven models, data-driven models typically do not consider the physical mechanisms underlying runoff formation. In recent years, with the rapid development of AI technologies, deep learning techniques such as Long Short-Term Memory (LSTM) (Hochreiter and Schmidhuber, 1997; Wu and Zhang, 2022; Wu *et al.*, 2022; Dong *et al.*, 2023), gated recurrent units (GRUs) (Wu *et al.*, 2023), and bidirectional LSTM (BiLSTM) (Tan *et al.*, 2023; Wu *et al.*, 2024) have been widely applied in runoff prediction and other fields. Dong *et al.* (2024) developed a runoff prediction model based on CEEMDAN-FE-VMD-SVM-GRU. First, the Complete Ensemble Empirical Mode Decomposition with Adaptive Noise (CEEMDAN) method was used to decompose the original daily average runoff dataset into multiple intrinsic mode functions (IMFs). Then, fuzzy entropy (FE) was introduced to calculate the complexity of each IMF component. IMF components with FE values greater than 0.4 were classified as high-frequency sequences, while those with FE values below 0.4 were classified as low-frequency sequences. Next, variational mode decomposition (VMD) was applied to further decompose the high-frequency sequences. Support vector machines (SVMs) and GRUs were used to predict the results of the first and second decompositions, respectively.

Yao *et al.* (2023) proposed a hybrid model based on CNN-LSTM and GRU-ISSA to predict runoff. This model also takes into account the influence of local meteorological and hydrological factors on runoff. First, the maximal information coefficient (MIC) was used to calculate the relationship between each variable and runoff, thereby reducing data dimensionality. Then, a convolutional neural network (CNN) was employed to extract runoff features from long time series data, and an LSTM network was used to make long-term runoff predictions. At the same time, a GRU was applied to predict runoff in shorter time series. Finally, an adaptive weighting module (AWM) was introduced to combine the advantages of both predictive models by dynamically learning their outputs and integrating them into the final prediction. To address the challenge of selecting model hyperparameters, an Improved Sparrow Search Algorithm (ISSA) was also employed.

Wu *et al.* (2023) also proposed a runoff prediction model based on VMD-CNN-AM-BOA-BiLSTM. First, VMD was used to decompose the original daily runoff sequence dataset into discrete IMFs and to distinguish signals of different frequencies. For each IMF, a CNN was introduced to extract its features. Then, a BiLSTM based on an attention mechanism (AM) was employed for prediction. Additionally, the Bayesian Optimization Algorithm (BOA) was used to optimize the BiLSTM's hyperparameters, further improving the predictive accuracy of the VMD-CNN-AM-BOA-BiLSTM model. In another study, Wu *et al.* (2021) developed a water demand prediction model for Taiyuan City based on principal component analysis (PCA) and backpropagation neural networks (BPs). Key indicators such as local population, regional GDP, precipitation, and average daily sunshine hours were selected, and PCA was used to reduce the dimensionality of the dataset and improve computational efficiency.

Finally, the BP neural network was used to predict water demand in Taiyuan.

2.4 Monitoring and Prediction of Water Environmental Disasters in the Yangtze River Delta Region

Given that water is an indispensable resource for human survival and social development (Ahmed *et al.*, 2019), accurately predicting water quality is crucial for ensuring the safety of drinking water and aquatic ecosystems. However, with intensified human activities, river water environments have significantly deteriorated, leading to frequent natural water disasters such as algal blooms (Jiang *et al.*, 2021). Ten water quality indicators—water temperature, pH, total nitrogen, ammonia nitrogen, total phosphorus, chlorophyll-a, CODCr, CODMn, dissolved oxygen, and total suspended solids—can be used for high-precision modeling of events such as red tide outbreaks and for issuing early warnings.

The permanganate index (CODMn) effectively reflects the extent of water pollution by reducing substances and is a key indicator for assessing such pollution. Improving the prediction accuracy of CODMn is crucial for aquatic ecosystem restoration and lake pollution prevention (Wang Yongshun and Cui Dongwen, 2024). In recent years, with the rapid development of computer technology, machine learning (ML) models, as a novel data-driven method, have been applied to CODMn or chemical oxygen demand (COD) prediction studies due to their strong nonlinear mapping abilities. Examples include BP neural networks (Zhu Lin *et al.*, 2024), BiLSTM networks (Chen Wei *et al.*, 2022; 2023), radial basis function (RBF) networks (Yang Zhuang *et al.*, 2017; 2019), adaptive neuro-fuzzy inference systems (ANFIS) (Pan Miao, 2019), least squares support vector machines (LSSVM) (Sun Jun *et al.*, 2017), quantum weighted minimal gating unit (QWMGU) (Zhang Yuze *et al.*, 2023), and SVMs (Li Jian *et al.*, 2023).

Dissolved oxygen (DO) is also a key water quality indicator that directly impacts the survival and reproduction of aquatic organisms (Liang *et al.*, 2023). Therefore, developing a DO prediction and early warning model capable of effectively monitoring and assessing river conditions is crucial. By predicting DO concentrations in advance, the aquatic ecosystem can be adjusted in a timely manner, such as by regulating water flow and oxygen circulation, to protect the aquatic environment and promote ecological balance (Azma *et al.*, 2023).

DO prediction methods can generally be divided into process-driven and ML-driven approaches (Wai *et al.*, 2022; Kerimoglu *et al.*, 2018). Process-driven DO prediction methods, such as the Soil and Water Assessment Tool (SWAT) (Noori *et al.*, 2020; Karki *et al.*, 2023) and the QUAL2KW model (Ghorbani *et al.*, 2022), simulate DO variations by constructing hydrodynamic models based on the current state and future time steps. While process-driven methods are powerful tools that clearly present the processes behind DO changes, they require comprehensive data on various factors such as topography, cross-sections, and other related parameters, limiting their applicability (Zhou, 2020).

Compared to traditional process-driven models, ML-driven models are widely used by scholars for DO prediction because they better capture the nonlinear relationships in complex data. Wu *et al.* (2024) proposed a knowledge-guided Catboost (KGCatboost) model for predicting DO concentrations. In addition, SHAP was used to analyze model interpretability, addressing the opacity of traditional ML models. Dong *et al.* (2018, 2023, 2024) proposed a model based on CEEMDAN-MFE-TVFEMD-SVM-LSTM. First, CEEMDAN was used to decompose the original DO data into multiple IMFs. Then, multiscale fuzzy entropy (MFE) was employed to calculate the entropy value of each IMF component. After linearly combining the high-frequency sequences, the Time-Varying Filter Empirical Mode Decomposition (TVFEMD) technique was applied to further decompose the combined high-frequency sequences. Finally, SVMs and LSTM neural networks were used to predict the sub-sequences from the first and second decompositions, respectively.

2.5 Prediction of Relative Sea Level Rise in the Yangtze River Delta Region

The IPCC reports suggest that regional mean sea level rise (RMSLR) is composed of global mean sea level rise (GMSLR) caused by global warming, tectonic subsidence (TS), and land subsidence (LS).

Accurately quantifying each of these components is the basis and prerequisite for all mitigation measures and represents a major challenge in international academia. Regarding GMSLR, the IPCC AR6 (2023) used climate models to simulate and predict historical and future global and regional sea levels, estimating a rise of approximately 0.43–0.84 m by the end of the 21st century. Wang *et al.* (2021), using satellite altimetry and global tide gauge data, accounting for natural variability and vertical crustal movement, projected a rise of 3.6–4.1 mm yr^{-1} by 2100. These values are significantly higher than the 2 mm/yr GMSLR adopted by the Shanghai government between 1996 and 2013, which was based on extensive fiscal and engineering measures to address global warming and sea level rise. This earlier estimate was calculated using data from the Wusong tide gauge station and covered mean tidal data trends from 1921 to 2000 (Cheng *et al.*, 2018; Cheng and Chen, 2017; Cheng Heqin and Chen Jiyu, 2016; Cheng Heqin *et al.*, 2015).

Regarding the extent of TS, the estuary region only has data from international joint measurements conducted by the Shanghai Astronomical Observatory using Very Long Baseline Interferometry (VLBI) from 1987 to 1993, and there have been no updates since. In recent years, several major earthquakes have occurred in East Asia (Webster *et al.*, 2005); therefore, their potential impact on TS in the eastern Chinese estuary region requires urgent evaluation. Furthermore, Professor Cheng Heqin's team at the State Key Laboratory of Estuarine and Coastal Research at East China Normal University analyzed data from 43 tide stations covering the Yangtze River estuary, from Jiujiang to Lüsi and Jinshanzui, and found that channel erosion and sedimentation caused by basin and estuarine engineering works significantly contributed to the regional mean sea level rise (RMSLR) (Cheng *et al.*, 2022; 2018; Cheng Heqin and Jiang Yuehua, 2021). Therefore, in light of the Paris Climate Agreement's goal of limiting global temperature rise to 1.5–2.0°C and China's established carbon neutrality targets, it is essential to establish infrastructure and conduct joint research using distributed multi-modal sensor data—including tide stations, satellite remote sensing, historical topography, and astronomical VLBI—to measure and address relative sea level rise.

Additionally, the project team analyzed data from the Mishi Ferry Station on the Huangpu River from 1970 to 2019, as well as precipitation data from the Taihu Basin, to study the trends, variability, frequency changes, and driving factors of annual maximum tide levels in the upper Huangpu River over the past 50 years. The results show that the annual maximum tide level has exhibited a significant rising trend over the past 50 years, increasing by 81 cm, with an average annual rise of 1.62 cm. After 1995, there was a sharp fluctuation and increase, with a large rise from 1996 to 2002 compared to pre-1995 levels. From 2003 to 2011, there was a noticeable decline compared to 1996–2002 levels, followed by another substantial increase. As tide levels rise, the corresponding maximum tide levels for the same frequency increase significantly, with the increase exceeding 0.9 m. The 50-year, 100-year, and 1,000-year recurrence interval tide levels increased from 3.77 m, 3.81 m, and 3.94 m in 1970 to 4.71 m, 4.75 m, and 4.92 m in 2019, respectively. Meanwhile, the recurrence intervals for the same tide levels drastically shortened, and the frequency of high tide events increased significantly. The rise in annual maximum tide levels is attributed to a combination of climate change and human activities. From 1970 to 1995, the rise was mainly influenced by climate factors, such as changes in precipitation, while from 1996 to 2019, it was primarily driven by factors such as water conservation projects and the increased frequency of typhoons, with the construction of projects such as the one on the Taipu River in the Taihu Basin playing a particularly important role. Therefore, the online monitoring, prediction, and prevention of sea level rise impacting the central urban area of Shanghai is of paramount importance and urgency for disaster prevention and mitigation in the city.

2.6 Fine-Scale Wind–Wave Field Inversion in Local Waters of the Yangtze River Delta under Extreme Weather Conditions

In recent years, as climate warming has progressed, typhoon tracks in the Western Pacific have shifted northward, with increasing frequency and intensity (Wu *et al.*, 2015). Furthermore, relative sea level rise has heightened flood levels in the Yangtze River Estuary, compounding the disaster impact of extreme weather events that cause extreme water levels and wave heights. This makes the consequences even more severe. As a result, more research has focused on the impact of sea level rise on coastal environments (Liu *et al.*, 2020). The IPCC Special Report on the Ocean and Cryosphere in a Changing Climate also highlights that the impact of sea level rise on coastal areas must consider the combined effects of gradual sea level rise and sudden extreme water level events, such as storm surges and wave action (Fischer and Knutti, 2016; Wu *et al.*, 2016). However, there is currently a lack of research on fine-scale wind and wave fields in local waters under extreme weather conditions, such as typhoons and cold waves, making it an urgent area for further study.

2.7 Intelligent Perception of Coastal Geomorphological Disaster Risks in Estuarine Deltas

Coastal geomorphological disasters and their evolution in estuarine deltas pose direct risks to flood protection, water supply, port navigation channels, and ecological safety. Examples include bank collapses, scour pits, steep slopes, estuarine sandbars, and sand waves. Their effective identification depends on high-resolution integrated land–water terrain monitoring equipment for shallow and extremely shallow waters, but no such equipment currently exists internationally. Moreover, there has been no research on intelligent identification technology for underwater disaster geomorphology, except for an exploratory study conducted by the project team. This study focused on the automatic identification of geomorphological elements using computer vision-based methods (Geomorphone) and self-organizing maps (SOMs), which were further improved. Given the combined impact of climate warming, accelerated sea level rise, reduced sediment supply from river basins, and large-scale land reclamation in estuarine regions, there is an urgent need to develop underwater disaster geomorphology monitoring equipment and intelligent risk identification technology.

2.8 Intelligent Identification and Early Warning of Flood and Tidal Defense Safety Risks in Estuarine Deltas

Sea level rise causes an elevation in storm surge levels (Zhang Jianyun and Xiang Yan, 2018), while also increasing nearshore wave heights and increasing local water depth leads to higher breaking wave heights in front of seawalls (Li Jialin *et al.*, 2006). This phenomenon is particularly evident in the Yangtze River Estuary, where the combination of floods, typhoons, and high tides (known as the "triple encounter or quadruple encounter") exerts significant pressure on flood protection and drainage systems in the region (Moftakhari *et al.*, 2017). Currently, research on the impact of relative sea level rise on coastal hydrodynamic disasters remains largely qualitative, with insufficient data and a lack of systematic quantitative analysis of the intensifying effects of such disasters. On the other hand, waves are the primary driving force behind seawall damage. Domestic and international researchers have conducted numerous physical model experiments to study the stability of toe and revetment stones (Jevrejeva *et al.*, 2019; Fang *et al.*, 2021). However, due to the complexity of seawall structures and their interaction with waves, there is no mature mathematical model to accurately simulate seawall stability under wave action. Additionally, there is a lack of systematic research on the failure mechanisms and early warning systems for seawall damage under conditions of relative sea level rise. Therefore, there is an urgent need for intelligent identification and early warning research on flood and tidal defense safety risks, particularly in the context of intensified extreme climate and weather events and the development of supporting infrastructure.

2.9 Intelligent Identification and Early Warning of Port and Navigation Channel Safety Risks in Estuarine Deltas

Under the combined effects of climate change and human activities, the rapid identification of macroscopic geomorphological hazards, such as estuarine bar formations, shoals, and submerged sandbars (Hou Yijun *et al.*, 2020), and microscopic hazards, such as sand waves, scour pits, and erosion channels (Chen Juzhong *et al.*, 2014; Li *et al.*, 2020; Clark *et al.*, 2018; Meehl *et al.*, 2012), will be crucial for ensuring the safety of global metropolitan international shipping hubs. Research using unsupervised learning algorithms to analyze historical measured big data for the intelligent identification and prediction of port and navigation channel geomorphological hazards has yet to be undertaken on an international scale.

2.10 Perception and Early Warning of Water Supply Safety Risks in Estuarine Deltas

The dynamics of estuaries are influenced by the combined effects of river flow, tidal currents, waves, extreme weather events, and topographical boundaries, making them exceptionally complex. The multi-level, braided systems of estuarine channels and shoals further enhance their three-dimensional complexity and make them highly sensitive to sea level rise (Schaeffer *et al.*, 2012; Nauels *et al.*, 2019). This sensitivity intensifies seawater intrusion, leading to urban and rural water supply shortages (Cheng Heqin and Chen Jiyu, 2016; Cheng Heqin *et al.*, 2015; Bamber *et al.*, 2019). To date, risk assessments and early warnings regarding the impact of climate warming and sea level rise on urban water supply safety in estuaries have included statistical analyses based on observational data (Schaeffer *et al.*, 2012; Rasmussen *et al.*, 2018; Jackson *et al.*, 2018), numerical simulations using models such as FVCOM, ECOM, and Mike (Qin Zenghao and Li Yongping, 1997; Syvistski *et al.*, 2009; Wang Hanmei and Jiao Xun, 2015; Shanghai Urban Planning and Land Resources Administration Bureau, 2013–2017), and one-dimensional analytical models that account for the influence of strong northerly winds (Pritchard *et al.*, 1956; Geyer *et al.*, 2014).

These results have provided technological support for water supply safety in the Yangtze River Estuary, but there is an urgent need to improve forecasting accuracy. The project team, leading a National Natural Science Foundation of China (NSFC) key international cooperation project, developed a vertically longitudinal 2D water and salt dynamics semi-analytical model for the Yangtze River Estuary using the iFlow model (Simpson *et al.*, 1990), which improved the accuracy of saltwater intrusion simulations. Additionally, the team used the ECOM model to calculate the impact of sea level rise on freshwater resources at the Yangtze River Estuary water source area from 2011 to 2030 (Oppenheimer *et al.*, 2019). They also employed the Vensim socio-economic system dynamics model to estimate changes in water demand over the same period (2011–2030). Based on critical climate change thresholds and corresponding adaptive strategies, the team developed a water supply safety model for the Yangtze River Estuary in Shanghai under different sea level rise scenarios for 2011–2030, including risk occurrence timelines, countermeasures, and an early warning system (Cheng Heqin *et al.*, 2015; Giddings *et al.*, 2011). However, further research is needed to develop a more intelligent and upgraded version of this system.

Chapter 3

The Construction Strategy for a Smart Water Network in the Yangtze River Delta Region Against the Backdrop of Normalization of Extreme Climate

Over the past century, influenced by both human activities and natural factors, global warming has become a common threat and challenge facing humanity. Extreme climates, such as heatwaves, heavy rains, floods, droughts, and wildfires, are becoming increasingly normalized. The United Nations issued a severe warning in 2023 that "the era of global boiling has arrived." The Yangtze River Delta region, which holds an important position in China's and the world's economic landscape, encompasses three provinces and one city (Shanghai, Jiangsu, Zhejiang, and Anhui) and is home to a permanent population of 237 million. The atmospheric, water, and soil environments within the region are closely interconnected, and the well-being of people's livelihoods across the area is interdependent. Against the backdrop of normalization of extreme climate, ensuring the water needs for production and daily life in the Yangtze River Delta region, especially for disaster prevention and reduction and for economic security, through the construction of a smart water network is a key measure to empower the region's high-quality, integrated development.

3.1 Current Situation and Problems in the Water Network of the Yangtze River Delta Region

3.1.1 *Poor connectivity of the water network in the Yangtze River Delta region*

The Yangtze River Delta encompasses the entire Yangtze River and Huaihe River basins, as well as the Taihu Lake basin and the southeastern regions of Zhejiang and Anhui. It features major rivers such as the Yangtze, Huaihe, and Qiantang, as well as renowned lakes such as Taihu, Chaohu, and Hongze, along with artificial waterways including the Beijing–Hangzhou Grand Canal, the Yangtze River to the Taihu River, and the Huaihe River flowing into the sea. It is one of the regions with the densest river networks in China and is also a typical coastal area consisting of a lowland plain water network. Influenced by the interplay of rivers, lakes, and marine environments, the region exhibits complex confluence characteristics, weak hydrodynamics, poor self-purification abilities of water bodies, and a tendency to accumulate pollutants. Moreover, since the 20th century, with the rapid advancement of urbanization and industrialization, large-scale activities have been undertaken in the Yangtze River Delta region, such as repairing embankments, building dams, and blocking rivers and gates, thus artificially severing the natural connections between rivers and lakes and between rivers and streams. This has resulted in poor connectivity of some water systems, a further decline in river network hydrodynamics, and serious encroachment on aquatic ecological space.

Taihu Lake, Chaohu Lake, and other important lakes in the Yangtze River Delta were originally part of river systems but have now been blocked; Nanyi Lake, Gucheng Lake, and other areas have also become closed or semi-closed water bodies due to the construction of gates, leading to significantly reduced water level fluctuations and weakened water exchange. Some small- and medium-sized river sections are almost stagnant during the non-flood season, making it difficult to ensure ecological base flow. These issues further reduce the carrying capacity of the water ecological environment in the plain water network, making the river and lake ecosystems more fragile.

Due to the fact that the rivers and lakes in the Yangtze River Delta are shared public goods among three provinces and one city, it is an urgent matter to adopt institutional constraints to internalize environmental externalities. Considering the significant differences in policy systems and regulatory standards among different regions, the current stage of smart water network construction faces challenges such as poor organizational coordination and institutional conflicts across various fields, including water resource monitoring information sharing, cross-border water pollution control, ecological environment co-protection, and cross-border ecological environment law enforcement. It is urgent to adopt the concept of "co-protection and joint governance" to clarify the water network governance responsibilities of different regions and levels and to enhance policy and institutional synergy in the construction process of smart and high-quality water networks in the Yangtze River Delta.

3.1.2 *Insufficient capacity for raw water configuration*

The Yangtze River Delta has essentially formed a network of water resources regulation through the construction of various water resources allocation projects. It relies on the Yangtze River, the Huaihe River, the Qiantang River, and other natural mainstream water sources, combined with Taihu Lake, Chaohu Lake, Hongze Lake, Gaoyou Lake, and Dianshan Lake, among other important water source lakes. However, there remain insufficient water resources and allocation projects in the northern regions of the Yangtze River, as well as inadequate water resources allocation management and scheduling capacity in coastal areas around Taihu Lake and eastern Zhejiang Province. Therefore, it is necessary to further optimize the network pattern of water resource allocation, strengthen research on large-scale raw water linkages, and enhance the configuration capacity of the water network.

The smart and high-quality water network in the Yangtze River Delta involves "one axis, one core, one screen, one belt, and three corridors with multiple points." The construction process requires multidimensional equilibrium regulation of ultra-large-scale water resource systems, supported by the theory of water resource equilibrium allocation and the dynamic regulation of multi-level water network engineering systems. The specific theoretical difficulties include incorporating multi-objective and multi-agent decision-making preferences from different regions and levels into the water resource allocation framework, integrating ecological protection cost control and ecological compensation into the framework, enhancing dynamic regulation of water resources under extreme climate uncertainty forecasting, and achieving cross-regional demand conflict coordination and win-win mechanisms under extreme climate conditions.

Moreover, as the economic leader of the Yangtze River Delta, Shanghai severely lacks local water resources, with all its four water sources relying on transit water from the Yangtze River and the Huangpu River. Among them, the Yangtze River's water source area is significantly affected by saltwater intrusion efforts resulting from Yangtze River runoff and storm surges. Although the water quality of the Huangpu River is generally above Class II, it is susceptible to sudden pollution due to its open nature. During his inspection in Shanghai, Xi Jinping emphasized the important mission of focusing on building five international centers for economic, financial, trade, shipping, and technological innovation. Shanghai's world-class strategic emerging industry clusters and evolving international business landscape are driving increasing demand for water resources and a favorable water environment. However, the contradiction between urban water consumption and the insufficient supply and quality of existing water sources is particularly apparent, especially under the context of normalized extreme climate and emergency scenarios. Water has become a crucial factor restricting the sustainable development of these five centers in Shanghai.

In 2022, during a rare drought in the watershed, the interception and discharge of water by tributaries increased, and the tidal boundary flowed 326 kilometers upstream from Jiangyin. Saltwater invaded the North Port South Branch and reached Dongfeng Xisha near Chongtou, severely restricting water supply to Yangtze River transit water sources such as Qingcaosha, Dongfeng Xisha, and Chenxing Reservoir.

On the other hand, the future water supply capacity of Shanghai is worrying. The "Shanghai Water Supply Plan 2019–2035" uses the research findings from major projects conducted by our team, which were undertaken by the Shanghai Science and Technology Commission between 2010 and 2013, to calculate the maximum unsuitable water intake days and assess the impact of sea level rise on the three primary water sources—Chen Hang, Qingcaosha, and Dongfeng Xisha—within the Yangtze River estuary. The research results indicate a relative sea level rise of 5–8 mm/year. Recent research by our team has shown that the frequency and intensity of typhoons and cold waves in the Yangtze River estuary area have increased over the past 10 years. The actual observation results from tide level stations, such as Wusong and Huangpu Park, indicate that the calculation of relative sea level rise in the Yangtze River estuary needs urgent correction, as well as its impact on the water source area.

3.1.3 *The pressure of flood discharge urgently needs to be relieved*

The Yangtze River Delta region has formed main channels for discharging floodwaters into the sea, with the Yangtze River, Huai River, and Qiantang River serving as primary outlets. The Huai River basin flows eastward into the sea and southward into the Yangtze River. The flood control measures in the Taihu Lake basin make full use of the lake for regulation and storage, discharging water northward into the Yangtze River, eastward into the Huangpu River, and southward into Hangzhou Bay. However, the layout of the flood discharge and drainage network for the southeast rivers flowing solely into the sea is incomplete. Due to siltation of flood discharge channels, the occupation of flood discharge and storage space, and dramatic changes in the underlying surface, issues such as the single outlet and insufficient scale of flood control in the Huaihe River, Taihu Lake, Chaohu Lake, and other important rivers and lakes are particularly prominent. The normalization of extreme climate and sea level rise will exacerbate these problems. Therefore, accelerating the construction of a smart and high-quality water network in the Yangtze River Delta will provide a more accurate basis and technical support for expanding flood discharge capacity and opening new flood channels, playing an important role in future disaster prevention and mitigation.

3.1.4 *The navigation capacity of integrated waterways in the Yangtze River Delta needs to be improved*

The Yangtze River Delta's high-grade waterway network is an important component of China's "two horizontal, one vertical, two networks, and eighteen lines" high-grade waterway layout. It has formed a "two vertical and six horizontal" waterway network pattern through the Beijing–Hangzhou Grand Canal, Hangzhou–Ningbo Grand Canal, Lianshen Line (including Yanglintang), Huaihe River Estuary Channel, Yanhe–Tongyang Line, Wushen Line, Suzhou–Shenzhen Outer Harbor Line, Changhushen Line, Huangpu River–Dapu Line, Zhaojiagou–Dalu Line (including Hujia Line), and Qiantang River–Hangzhou Shen Line (including Hangzhou Pingshen Line). However, with the rapid economic and social development in the Yangtze River Delta region, the original waterway layout can no longer fully meet the water transportation needs of the three provinces and one city. The navigation capacity, accessibility, and connectivity of high-grade waterways need to be enhanced, particularly in terms of docking with coastal ports and cross-provincial and cross-municipal waterway connections, which still pose certain challenges. The construction of a smart and high-quality water network in the Yangtze River Delta will provide important guarantees for optimizing and improving the region's waterway network and enhancing cross-regional and cross-disciplinary transportation channel capabilities.

3.2 The Importance and Guiding Direction of Building a Smart and High-Quality Water Network in the Yangtze River Delta

The Intergovernmental Panel on Climate Change (IPCC) released its report, "AR6 Synthesis Report: Climate Change," in 2023, clearly stating that human activities have caused warming of the atmosphere, oceans, and land. Based on a comprehensive assessment of multiple variables within the climate system, the report found that the global surface temperature from 2011 to 2020 was 1.1°C higher than that from 1850 to 1900, directly resulting in extreme weather events such as high temperatures, droughts, severe flooding, storms, and decreased crop yields worldwide. Even if the global temperature rise is limited to 1.5°C, the proportion of people threatened by water scarcity, heat stress, and flooding will still significantly increase. In this context, the Yangtze River Delta region urgently needs to rely on the "National Water Network Construction Plan Outline" to accelerate the construction of a smart and high-quality water network and take proactive measures to respond and adapt to the challenges posed by the normalization of extreme weather. It should be noted that, compared to traditional water system networks, the smart and high-quality water network in the Yangtze River Delta is not merely a simple engineering network but a comprehensive system based on natural rivers and lakes. It includes water diversion and drainage projects as channels, regulation and storage projects as nodes, and intelligent regulation as a means. This system integrates functions such as the optimization of water resource allocation, flood control and disaster reduction in river basins, and the protection of water ecosystems. It holds significant strategic importance in ensuring regional ecological balance and national economic security.

High-quality water resources are not only a crucial production factor supporting the development of industries in the Yangtze River Delta region, but they are also the most important public good concerning the vital interests of the public in the region. Their supply, maintenance, and governance exhibit typical environmental externalities, involving the distribution of interests and ecological compensation between upstream and downstream areas, main rivers and tributaries, and the left and right banks. Given the context of normalized extreme weather, if we approach the allocation of water resources in different cities within the Yangtze River Delta with a localized, fragmented, and self-interested mindset, we will inevitably fall into the dilemma of "everyone managing their own affairs and mutually constraining each other." Since the 18th National Congress of the Communist Party of China, General Secretary Xi Jinping has proposed the concept of "giving priority to water conservation, maintaining spatial balance, pursuing systematic governance, and implementing dual-track efforts" in water management. At the 2023 Symposium on Deepening the Integration of the Yangtze River Delta, he emphasized the need to "strengthen the seamless connection of ecological protection red lines among the three provinces and one city, and deepen the joint protection and governance of cross-border water bodies." The aim is to guide the construction of a smart and high-quality water network in the Yangtze River Delta, incorporating a systemic approach that breaks down regional administrative barriers and ensures the network is "complete, safe, and reliable; intensive, efficient, green, and intelligent; with smooth circulation and orderly regulation."

3.3 Benefit Evaluation of Smart and High-Quality Water Network Construction in the Yangtze River Delta

The high-quality, integrated development of the Yangtze River Delta must closely adhere to the two key principles: integration and high quality. To build a smart and high-quality water network in the Yangtze River Delta, we will rely on an intelligent perception network that includes all elements, with suitable spatial and temporal density, and integrates the sky, earth, and sea. This network will break the boundaries of river basins, regions, and water sectors, coordinate the comprehensive governance systems of the Yangtze River, Huaihe River, Taihu Lake basins, and southeast rivers, and address the demand for water security assurance across provinces and cities. It will promote the transformation of water resource allocation in the Yangtze River Delta to be more systematic, proactive, and refined, strengthen existing hydrological monitoring equipment, reduce flood disaster losses, and generate significant economic and social benefits.

From the perspective of transforming ecological value into economic value, building a safe, high-quality, interconnected, and mutually beneficial smart water network in the Yangtze River Delta, and achieving regional integrated water resource allocation can not only address spatial imbalances in water resources and local water quality shortages but also improve the quality of raw water in the Yangtze River Delta urban agglomeration and reduce the cost of water purification in megacities such as Shanghai, Nanjing, Hangzhou, and Hefei. In addition, adhering to the concept of "green mountains are as valuable as mountains of gold and silver," the environmental well-being of the public in the Yangtze River Delta region can be improved by relying on smart and high-quality water networks to create a harmonious river and lake network that prioritizes ecology and promotes harmony between people and water and restores the ecological quality of river networks and lakes. It can also promote the construction of world-class natural ecological and cultural tourism corridors in the region and provide new momentum for the ecological and tourism industries in the Yangtze River Delta.

From the perspective of coordinating high-quality development and high-level security, building a smart, high-quality water network in the Yangtze River Delta will effectively address the problems of increased water demand, increased difficulty in guaranteeing water supply, and insufficient emergency response capabilities brought about by urbanization and industrial transformation in the region. It will strengthen the supporting role of water resources in modern agriculture, advanced manufacturing, and modern service industries within the Yangtze River Delta and fully empower the region's economic and social development. Through the smart and high-quality water network of the Middle and Lower Yangtze Valley Plain, we can provide high-quality water resources and strengthen the region's resilience and confidence in coping with extreme weather, such as high temperatures, droughts, and floods. This will also help stabilize the production and supply of grains, oilseed crops, tea, mulberry, and aquatic products in the middle and lower reaches of the Yangtze River Plain and provide important guarantees for the economic security of the Yangtze River Delta as well as the whole country.

3.4 Governance Difficulties in the Construction of Smart and High-Quality Water Network in the Yangtze River Delta

3.4.1 *Difficulties in regional coordination*

The construction of smart, high-quality water networks in the Yangtze River Delta necessitates comprehensive coordination of the integrated management systems of the Yangtze River, Huaihe River, Taihu Lake basins, and southeast rivers. Emphasis should be placed on connectivity, joint prevention and control, joint construction and management, and cooperation and coordination. Since the rivers and lakes in the Yangtze River Delta are public goods shared among three provinces and one city, it is imperative to adopt institutional constraints to internalize environmental externalities. Given the substantial differences in policy systems and regulatory standards across different regions, the current stage of smart water network construction encounters challenges such as organizational coordination and institutional conflicts in various areas, including water resource monitoring information sharing, cross-border water pollution control, ecological environment co-protection, and cross-border ecological environment law enforcement. It is crucial to adopt the concept of "co-protection and joint governance" to clarify the water network governance responsibilities of different regions and levels, thereby enhancing policy and institutional synergy in the construction process of smart and high-quality water networks in the Yangtze River Delta.

3.4.2 *Difficulties in resource allocation theory*

The smart and high-quality water network in the Yangtze River Delta encompasses "one axis, one core, one screen, one belt, and three corridors with multiple points." The construction process necessitates multidimensional equilibrium regulation of ultra-large-scale water resource systems, which requires support from theory regarding water resource equilibrium allocation and the dynamic regulation of multi-level water network engineering systems. The specific theoretical challenges include incorporating multi-objective and multi-agent decision-making preferences from different regions and levels into the water resource allocation framework, integrating ecological protection cost control and ecological protection compensation into the water resource allocation framework, enhancing dynamic regulation of water resources in the context of extreme climate uncertainty forecasting, and achieving cross-regional demand conflict coordination and win-win mechanisms under extreme climate conditions.

3.4.3 *Key technological innovation difficulties*

Building a smart and high-quality water network in the Yangtze River Delta, amidst the normalization of extreme weather conditions, will pose greater technical challenges compared to traditional water network construction. For instance, in key technical domains such as extreme climate change monitoring systems based on remote sensing, artificial intelligence, and digital simulation, water quality and hydrological monitoring sensing systems, large-scale water diversion project joint debugging and control models, digital twin water network and physical water network synchronization simulation platforms, and water network tunnel construction under complex geological conditions, there exist numerous technological innovation blind spots and significant equipment dependency issues. It is imperative to integrate talent, technology, and other innovative resources from various regions and, through collaborative research and development, overcome the technical barriers to constructing smart and high-quality water networks.

Chapter 4

Intelligent Water Network Management Platform for the Yangtze River Delta: A Construction Scheme in the Context of Extreme Climate Normalization

In the context of extreme climate normalization, the construction of an intelligent and high-quality water network management platform in the Yangtze River Delta needs to focus on addressing the following key issues:

(1) **Impact mechanisms and assessment methods:** Research the impact mechanisms and assessment methods of extreme climate events on water supply–demand balance, water environment quality, and ecosystems in the Yangtze River Delta. Establish adaptive water resource allocation and safeguard plans that adapt to extreme climate changes, and improve water resource utilization efficiency and conservation capacity.

(2) **Spatial and dynamic characteristics:** Study the spatial distribution, hydrological and hydraulic characteristics, and dynamic changes in water quantity and quality within the water network of the Yangtze River Delta. Build an intelligent and high-quality water network management platform based on the Internet of Things (IoT), cloud computing, big data, artificial intelligence (AI), and other technologies to achieve dynamic

monitoring, accurate prediction, timely warning, and scientific management of the water network.

(3) **Emergency response mechanisms:** Investigate emergency response mechanisms for the water network management platform in the Yangtze River Delta and formulate plans to address water-related disasters, such as floods and droughts, caused by extreme climate events. Establish a risk assessment and prevention system based on multi-scale weather forecasts and impact predictions, enhancing disaster resilience and water disaster prevention and response capabilities.

Focusing on these areas, the current project has conducted extensive literature research in the early stages, collecting and analyzing relevant domestic and international policies, standards, technical manuals, scientific papers, and expert reports to understand the background and current status of intelligent, high-quality water network construction. Additionally, through field investigations, personal interviews, and academic exchanges, the project has explored cutting-edge technologies and advanced experiences, gained insight into local policies and management systems, and understood the water usage needs of relevant enterprises and residents, as well as the use of water infrastructure. This provides a solid data foundation for subsequent research.

4.1 Current Situation

The construction of Digital China and the digital economy is currently a core strategy for our country and will together drive the digital transformation of the water sector, enabling high-quality development of intelligent water management. On May 25, 2023, the Central Committee of the Communist Party of China and the State Council issued the "National Water Network Construction Outline." This initiative aims to establish key national water network projects by 2025, accelerate the construction of the national backbone network, and improve the intelligence of water network projects. By 2035, the overall framework of the national water network will be largely formed, with the main framework and major arteries of the national water network gradually completed and provincial, municipal, and county-level water networks nearly perfected. This will establish a national water security system aligned with the fundamental realization of socialist modernization.

The planning outline highlights three key points for "accelerating intelligent development":

(1) **Digital construction of the water network:** Strengthen the digital construction of the water network, deepen the integration of national water network projects with new infrastructure construction, and promote the digital and intelligent development of water network projects.
(2) **Intelligent dispatch management:** Enhance the intelligence of water network dispatch management, accelerate the construction of the National Water Network Dispatch Center, Big Data Center, and Basin Sub-centers, and establish a national water network dispatch and command system.
(3) **Monitoring system improvement:** Improve the water network monitoring system by applying new technologies and methods, such as next-generation communication technologies, high-resolution remote sensing satellites, and AI systems. Enhance the automation and intelligence of monitoring equipment, and create a comprehensive, high-precision, multidimensional, and secure water network monitoring system.

Given its location at the estuary of the Yangtze River, the water resources and environment in the Yangtze River Delta are significantly impacted by challenges such as seawater intrusion. In the context of the normalization of extreme climate, water resource problems cannot be ignored. Therefore, it is crucial to integrate smart water supply and usage, and building an intelligent, high-quality water network in the Yangtze River Delta is imperative.

The intelligent water network can rely on IoT monitoring and AI to closely connect water, users, and the water treatment end. It enables real-time monitoring, analysis, and control in the drainage system, from rainfall to urban rivers. Building an intelligent, high-quality water network platform in the Yangtze River Delta can achieve automatic monitoring, flood prevention, and disaster reduction; reduce the need for on-site personnel; promote data sharing; and facilitate regional water resource integration. This will ensure the sustainable development of urban water resources and contribute to the achievement of urban modernization goals.

The concept of the "air–space–ground–sea" integrated network was first proposed in the field of communications. This integrated information network serves as an information infrastructure that extends ground networks through space-based networks, encompassing natural environments such as outer space, the atmosphere, land, and oceans. It provides information support for the activities of various users, including space-based, air-based, land-based, and sea-based platforms. The development and construction of such a network not only reflect a nation's comprehensive technological and economic strength but also serve as a key driver for promoting independent innovation in major sectors, meeting national strategic needs, and enhancing the country's competitiveness in cyberspace. Since the 21st century, China has actively pursued the construction of integrated information networks, striving for control over cyberspace, air, and sea, while promoting the integration of space-based networks with ground-based and mobile communication networks. Several space-based networks have already been established, including civilian systems such as Starlink and OneWeb, military systems such as the Advanced Extremely High Frequency (AEHF) satellite communication system, and dual-use systems such as the Iridium Next system. Overall, these networks demonstrate increasingly mature information transmission technologies, expanding information capacity, diverse network architectures, and extensive service coverage.

Due to the scarcity of water resources, there has been a growing demand in recent years for higher levels of automation, informatization, and visualization in the water industry. The intelligent water network is a water management model that uses information technology, including IoT and big data, to intelligently and automatically monitor, dispatch, manage, and utilize water resources systematically. The intelligent, high-quality water network management platform is such a smart water system. Through the air–space–ground–sea IoT module, it monitors and collects massive amounts of information from water-related equipment (such as on-site data collectors, water quality and pressure meters, and other online sensing devices). This includes data on municipal water supply, industrial water usage, wastewater treatment, estuary water salinity, suspended sediment concentration, seawater desalination, water source monitoring, municipal water supply and drainage, rainwater recycling, and water supply equipment information.

Various equipment statuses and data are collected and aggregated on cloud servers, enabling the platform to analyze and process massive amounts of water system information in a timely manner. This facilitates more precise and dynamic management of the entire production, management, and service processes of the water system, improving production efficiency and reducing operational and maintenance costs, ultimately achieving a state of high efficiency, capability, and intelligence. Through the air–space–ground–sea IoT module, all connected devices can be monitored and information can be retrieved in real time. A browser or a mobile app client can display the real-time operating conditions of all connected devices, showing the obtained real-time data laterally (among different devices) and longitudinally (historical, from a single device). Simultaneously, the module performs information management on all networked water treatment equipment, ensuring a full record of equipment maintenance.

The intelligent water network is an urban water supply infrastructure enhanced by various interconnected devices, with the capability to collect and share data with other devices and data centers. This is typically achieved through the IoT technology. Both domestic and international scholars have conducted extensive research on this and proposed their own solutions:

(1) **Real-time hydrological monitoring:** IoT technology, by deploying sensors at key water resource points, can collect real-time data on important hydrological parameters such as water levels and flow rates. This real-time data collection mechanism provides unprecedented transparency and immediacy for water resource management. Compared to traditional manual monitoring, IoT technology significantly reduces data collection time and labor costs while improving data accuracy and reliability. This real-time monitoring capability allows managers to stay informed about the latest water resource conditions and respond quickly to potential changes or issues.
(2) **Real-time water quality monitoring:** Traditional methods require collecting samples from the source, transporting them,

preserving them, and analyzing them in a laboratory. This process is not only time-consuming and labor-intensive but also susceptible to changes in sample composition due to environmental factors during transportation, which can lead to decreased accuracy. To overcome the issues with traditional water quality testing, the U.S. Geological Survey proposed a method for continuous monitoring of water quality. With the development of communication technologies (mainly wireless sensor networks (WSNs) and the IoT), the range of monitored key parameters has expanded to include pH, turbidity, dissolved oxygen, microorganisms, and more.

(3) **Water supply network monitoring:** Intelligent pipes are modular units with monitoring capabilities, allowing for continuous expansion and updates to their sensors. By installing smart pipes at key sections of the water supply network, near real-time monitoring of flow, pressure, water quality, and leakages can be achieved without disrupting the operational efficiency of the pipelines. The collected data are transmitted through a wireless sensor network, with the main advantage being the ability to continuously monitor without the need for local manual intervention and the low energy consumption of wireless sensors ensuring long-term continuous operation.

(4) **Intelligent water meter:** Traditional water meters are devices used to measure water consumption. Intelligent water meters offer additional features, such as storing and transmitting water usage data. In a smart water network, bidirectional communication between water utilities and water meters is necessary, and communication with customers may also be established. Although traditional water meters can also have their usage data manually read, smart water meters can provide long-distance, high-frequency data on water consumption, offering users and management entities instant access to near real-time information. Thus, intelligent water meters present an opportunity to improve drinking water supply, enable management entities to effectively collect service fees, and foster a shared responsibility to protect water as a scarce resource.

According to relevant surveys, over 95% of Chinese water utility companies have closely followed industry trends in smart water management, and an increasing number of digitally advanced water companies are initiating top-level design projects. More than 80% of water utilities recognize the value of smart water management by providing substantial support in terms of organizational structure, financial investment, policy formulation, and human resource allocation. Over 60% of companies have expressed strong support for intelligent water management, including substantial support in terms of organizational structure, financial investment, policy formulation, and human resource allocation. Additionally, 65% of water utilities are focused on infrastructure development, 52% are concerned with information systems coverage, 56% are focused on integration, and more than 60% are interested in the utilization and exploration of data resources.

With the development and rise of intelligent water management, significant transformations have occurred within the digital management departments of relevant water utility companies. According to data from the "Smart Water Management Survey and Analysis Report (2020)" conducted by the China Water Association, 50% of the surveyed water companies have established independent digital management departments, with some even forming dedicated leadership teams for intelligent water management. This indicates that China's intelligent water systems have made rapid progress in terms of standardized construction, talent reserves, technical support, and scientific research achievements. However, when it comes to setting standards for intelligent water management, existing local standards are predominantly concentrated in municipalities and eastern regions, resulting in an imbalance in regional development. Linear group standards involve multiple associations and focus on areas related to the scope of each association's activities, while there are fewer standards related to the application of intelligent water business.

Compared to the digital water management in Europe and America, which emphasizes theoretical innovation and fundamental methodologies, China's intelligent water grid research emphasizes technology application and solving practical water management issues, such as water quality monitoring and pollution control. This is mainly achieved through collaboration between government agencies, research institutes, universities, and enterprises. There are a total of 95 national-level projects using the keywords "intelligent + water", with a higher funding rate for city-related topics. However, many of the funded areas are highly interdisciplinary and only loosely connected to water management, with relatively low involvement from water management researchers. One notable achievement is the independently developed Simuwater platform by the North China Institute of Water Conservancy. As the first domestic hybrid model, it integrates mechanistic models, generalized models, and AI, enabling seamless simulation of urban drainage systems at the watershed level. This innovation overcomes the limitations of foreign-controlled drainage system simulations. Meanwhile, Harbin Institute of Technology has led the development of an AI-based carbon emission accounting model (Matougang Wastewater Plant Project). By combining mechanistic models and data-driven algorithms, a "carbon metabolism" model for biological treatment processes was established, developing a real-time carbon emission metering method and achieving minute-level carbon emission measurement. Another significant project, the Boai No. 2 Wastewater Treatment Plant project, employs digital twin technology to enable low-carbon operations. This involves the use of 3D visualization modeling, process simulation, automatic control simulation, and 3D visualization techniques to build a 3D virtual water plant. It integrates static equipment data with dynamic operational data, reflecting on-site details in real-time and realizing functions such as water quality prediction, water quantity forecasting, biochemical analysis, material balance, operational adjustments, equipment maintenance, and fault diagnosis.

In addition to AI applications, a wide range of digital products have emerged in China's water management industry, such as smart manhole covers, smart water pumps, smart valves, smart hydrants, hydraulic control equipment, and grid-based water accumulation identification systems. These digital products are crucial prerequisites for obtaining truly real-time and effective big data, forming a critical foundation for data integration into a broader data asset framework. By utilizing these digital products, it is possible to quickly respond to specific national and urban needs as well as new challenges.

Although China's intelligent water network is currently developing rapidly and positively, several issues remain:

(1) **Integration of digital technologies and business applications:** While applications and business operations are precise, there is a lack of knowledge about some emerging technologies. Professional technical personnel from universities and research

institutes possess cutting-edge scientific expertise, but their understanding of practical application scenarios and business needs remains insufficient.

(2) **Measuring benefits:** The benefits of intelligent water networks are challenging to quantify. The smart water network itself does not easily generate new economic benefits, and the specific benefits of digital projects are difficult to quantify.

Moving forward, the focus will need to shift toward better addressing the technical challenges of integrating intelligent equipment and industrial software with water treatment processes. It will require a transformation in thinking, the integration of digital technology with specific business applications, the integration of big data and artificial intelligence, security assurance, and the cultivation of professional talent, which will be the top priorities for the future development of smart water networks.

4.2 Plan for Building an Intelligent and High-Quality Water Network Management Platform for the Yangtze River Delta in the Context of Normalized Extreme Climate

High-quality water is an essential necessity for people's lives. In recent years, the effects of global climate change and human activities on water resources have intensified, leading to increasing water shortages and pollution issues. Factors such as extreme weather, residential water waste, industrial discharge, saltwater intrusion, wastewater treatment efficiency, rainwater recycling, and water supply strategies and efficiency all play a critical role in the rational use of water resources. It is vital to employ digital technologies for real-time monitoring and management of water quality and quantity, enabling timely detection of anomalies and facilitating swift responses and adjustments.

4.2.1 *Construction of the air–space–ground–sea IoT network*

In response to current development needs for the intelligent water network, it is urgent to build an integrated "air–space–ground–sea"

Table 4.1. The composition and functions of the air–space–land–sea Internet of Things.

Sub-network	Composition	Function
Air-based network	Drones, airships, balloons, ground stations, etc.	Information collection, emergency rescue, temporary communication.
Space-based network	Geostationary satellites, low-Earth-orbit satellites, etc.	Navigation and positioning, information collection, communication relay.
Ground-based network	Mobile communication systems, communication base stations, ground sensor networks, etc.	Network communication, information collection, sub-network status monitoring, information processing and sharing.
Sea-based network	Buoys, vessels, base stations, underwater detection equipment, etc.	Information collection, navigation and positioning, maritime communication.

IoT system that applies IoT technologies to water supply equipment, wastewater treatment facilities, water pump gate groups, and smart irrigation systems. This study establishes a ground-based monitoring network integrating satellite monitoring, drone monitoring, and marine monitoring networks to achieve a fully comprehensive IoT system, as shown in Table 4.1.

In the air–space–ground–sea framework, the "air" component refers to drone monitoring, which provides refined remote surveillance over a certain area. Compared to the wide-range, non-differentiated monitoring of satellite remote sensing, drone monitoring is more targeted and can execute precise surveillance based on specific requirements. The drones receive commands through a wireless control system to perform operations, and the data they collect are transmitted back to the base station via 5G or Ethernet and then sent to the control system terminal—the central control system.

In the air–space–ground–sea framework, the "space" component refers to satellites equipped with meteorological monitoring, remote sensing, navigation, and communication functions. These satellites work collaboratively in a heterogeneous manner, covering extensive areas. They enable high-precision positioning in the upper atmosphere, on the sea surface, and in nearshore environments, enhance marine information collection, and act as communication relays between sub-networks.

In the air–space–ground–sea framework, the "ground" component consists of mobile communication systems, comprehensive communication base stations, and ground sensor networks. The mobile communication system supports the overall communication operations of the ground network, ensuring low latency and high reliability for stable communication and data privacy. The integrated communication base stations are responsible for interfacing with space-based, air-based, and sea-based networks, monitoring the status of subnet nodes, transmitting operational commands, and consolidating marine information. Ground sensor network nodes, including automatic monitoring stations and environmental sensors, collect and supplement nearshore marine data.

In the air–space–ground–sea framework, the "sea" component refers to the close monitoring of ocean and estuarine environments and water quality through methods such as deploying acoustic detection equipment on unmanned boats, installing observation instruments at river estuaries, and using submerged buoys for detection. These devices also transmit real-time data on water quality, salinity, suspended sediment concentration, sea level height, tidal conditions, and other information back to the base station via 5G or Ethernet, which are then passed to the control system terminal of the central control system.

The challenges of integrated construction of the air–space–ground–sea IoT are multifaceted. In response to the major issues identified, we propose the following countermeasures:

(1) **Difficulty in coordinating and managing information across multiple platforms:** At the current stage, each detection unit develops independently, with no standardized data formats, resulting in a wide range of heterogeneous and multi-source information. This creates significant difficulties in management and allocation. To overcome this, a unified data management standard can be established to store and manage data in a consistent format. Alternatively, efficient data transformation methods can be implemented to streamline the standardization of data formats, improving overall integration and management.

(2) **Data security is difficult to ensure:** The monitoring data involves maritime commercial data such as shipping and fisheries, along with scientific research data related to marine ecology and climate. Unauthorized access or data breaches could threaten national security, the marine economy, and scientific advancements. Therefore, data confidentiality is a crucial prerequisite for construction. For each subnet, security measures should be tailored to its specific characteristics, and a robust security mechanism for transmitting information between sub-networks must be implemented to safeguard data integrity.

(3) **Infrastructure urgently needs upgrading:** In international research, a significant focus is placed on infrastructure, particularly in areas such as intelligent water meters and intelligent pipeline networks. These types of sensors enable real-time and rapid monitoring and transmission of information, facilitating the ultimate goal of constructing intelligent water networks. However, China's infrastructure still has substantial room for improvement. For instance, many water meters rely on outdated methods such as water cards for reading and writing data, causing delays in the statistical analysis of water usage. In reservoirs and water transmission networks, there is a lack of corresponding sensors for monitoring, or sensors are present but the data are used internally and not made public. These issues lead to data gaps or inefficient application. Therefore, it is essential to assess the current state of infrastructure, protect and standardize the management of existing systems, and prioritize the construction or upgrading of outdated or insufficient facilities.

(4) **Limitations of existing communication networks:** As technology continues to advance and detection methods diversify, the air–space–ground–sea IoT will integrate an increasing number of specialized devices. To support real-time monitoring, communication methods will face higher demands for speed and efficiency. The limited coverage, low traffic density, and restricted positioning accuracy of 5G networks are insufficient to meet the evolving needs of IoT infrastructure. Therefore, a more reliable, low-latency, and highly stable communication solution—6G network technology—is required. Compared to 5G technology, 6G offers significant improvements in coverage, cost efficiency, and intelligence. Moreover, developing 6G networks based on the

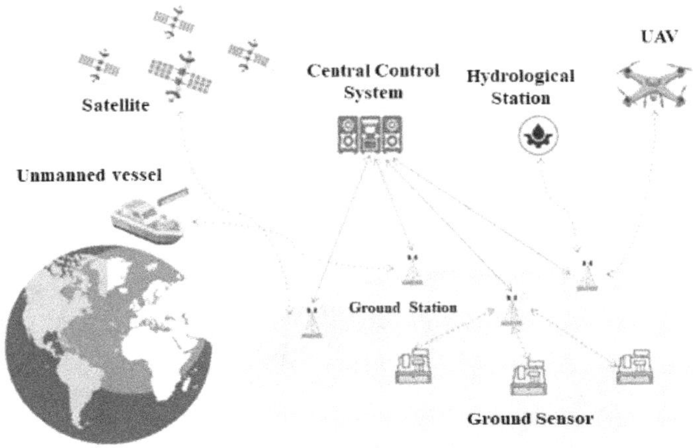

Figure 4.1. The control framework for the integrated air–space–land–sea network.

existing 5G framework will be crucial for meeting the more rigorous network requirements of future IoT applications.

In response to the current issues and corresponding solutions, the ultimate established air–space–ground–sea IoT control architecture is illustrated in Figures 4.1 and 4.2.

4.2.2 *Continuous water quantity and quality monitoring system with sensor placement*

A key element of an intelligent water network is the continuous monitoring of on-site water quantity and quality parameters. Critical parameters include pH, turbidity, dissolved oxygen, residual chlorine, total dissolved solids (TDS), and microbial contamination. Among these, a key parameter that must be monitored in real time is the residual chlorine concentration. Next, taking residual chlorine as an example, we specifically discuss how to detect these key parameters. Residual chlorine, also known as free chlorine, refers to the minimum chlorine level remaining in water after a period of use or contact. During transport from the treatment plant to the user, residual chlorine is an important safeguard against microbial contamination. It is associated with the

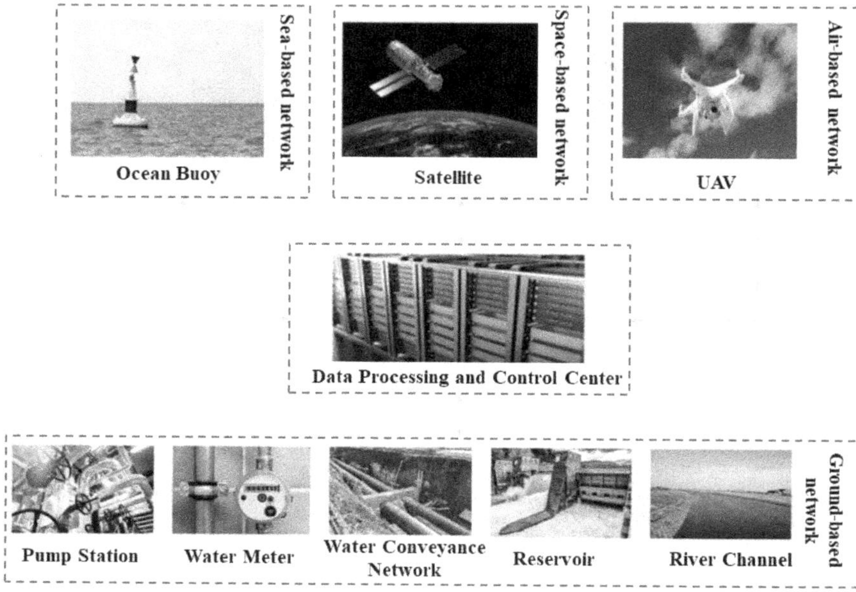

Figure 4.2. A conceptual construction diagram of the integrated air–space–land–sea network.

absence of microorganisms and is a measure of water's potability. Currently, a combination of mobile and fixed sensors, including ultrasonic level sensors and residual chlorine sensors, is used to monitor water tank levels, chlorine liquid levels, and chlorine concentrations in reservoirs. Real-time monitoring of chlorine concentration enables precise control of dosing pumps, with data transmitted to the cloud via WSNs and the IoT. These critical parameters require on-site water quality monitoring, which relies on WSN and IoT technology.

WSNs are a key technology within the IoT, consisting of numerous interconnected sensors. However, these sensors cannot transmit data directly to the internet. To facilitate this, intermediary devices such as routers or sink nodes are required. A typical IoT architecture uses gateway devices as a bridge between sensors and the cloud platform, communicating via various network protocols to transfer data to the cloud for storage. The cloud then handles the storage, analysis, and processing of the data, enabling more informed decision-making through real-time mobile data access and analysis. Sensor placement plays a crucial role in smart water networks. A commonly used objective function for optimizing sensor locations is to minimize detection time, impact, exposure scale, and the extent of damage. Proper placement improves the diagnostics of the water pipeline system, enhancing efficiency and reducing overall operational costs.

Chapter 5

Development of a Risk Exceedance Perception and Analysis System for Smart and High-Quality Water Network Infrastructure in the Yangtze River Delta

At present, the intelligent collection, processing, and analysis of basic information related to China's independent intellectual property rights in smart and high-quality water networks, along with capacity building for regional independent water resource solutions, constitutes the foundation and prerequisites. On November 30, 2023, General Secretary Xi Jinping presided over a symposium on deepening the integration of the Yangtze River Delta in Shanghai, explicitly calling for the enhancement of seamless connections among ecological protection red lines across the three provinces and one city, the promotion of joint protection for important ecological barriers and corridors, the strengthening of comprehensive prevention and control of air, water, and soil pollution, and the deepening of collaborative protection and treatment efforts for cross-border water bodies. In response to this directive and the normalization of extreme climate conditions, Shanghai and the surrounding areas urgently need to develop the "Yangtze River Delta Smart High-Quality Water Network Risk Exceedance Perception and Analysis System." This initiative will focus on the theories, methods, and technologies involved in the construction and operation of a smart, high-quality water network in the Yangtze River Delta region. It will encompass the monitoring and evaluation of the smart and high-quality water network, as well as multi-indicator balancing and synergistic effect analysis. Furthermore, it will entail the development of a risk perception and analysis system tailored to the smart and high-quality water network in the Yangtze River Delta. This system will involve the assessment of the current state of water resources, future demand projections, water network monitoring, equipment development, risk assessments, platform construction, and action plans.

5.1 Data Acquisition and Representation of the Smart and High-Quality Water Network in the Yangtze River Delta

5.1.1 *Online monitoring and prediction of physical elements in the Yangtze River Delta water network*

Based on a framework of intelligent and high-quality water network indicators, comprehensive utilization of domestic and international Earth big data sources (including satellite observations, economic and social data and statistical data) is conducted. Typical case studies of smart and high-quality water networks in the Yangtze River Delta are carried out at various scales, including regional, provincial, municipal, and county levels. A basic dataset and algorithm toolkit with independent intellectual property rights, which is publicly shareable and capable of meeting the needs of indicator monitoring and evaluation, is developed. This toolkit aims to enhance construction standards for key hydrological station networks and provide data, methods, and tools to support online monitoring and prediction of the spatiotemporal distribution of water resources in the Yangtze River Delta.

5.1.2 *Building a multi-scale extreme weather and climate event prediction and warning system to enhance the intelligent water network's ability to perceive exceedances*

Ensuring water network security necessitates the ability to reliably predict the occurrence of extreme events in the coming days and anticipate potential changes in extreme events over the coming years to decades. Short-term extreme event prediction and medium- to long-term extreme climate change estimation each pose distinct scientific and technological challenges. To make short-term predictions, on the one hand, it is crucial to strengthen the detection of various extreme event generation factors and utilize satellite, radar, flow, and tidal intelligent monitoring systems to establish a high-density, high-quality monitoring network. This will provide comprehensive data, information, and support for predicting recent extreme events. On the other hand, there is a need to develop the predictive capabilities of artificial intelligence meteorological models for extreme events. In terms of prediction efficiency, artificial intelligence meteorological models offer advantages over traditional physics-based prediction models and can better meet the requirements for short-term forecasting of extreme events, ensuring timely predictions. The spatio-temporal high-density monitoring system for extreme event generation factors can provide more comprehensive training data for meteorological large-scale model training. For medium- to long-term estimates of extreme event changes, it is first necessary to strengthen scientific research on the sources of uncertainty in such estimates for the Yangtze River Delta region. This involves clarifying the contributions of natural variability and external forcing within the climate system to the uncertainty of estimates across different future periods, thereby improving the reliability of those estimates in a targeted manner. On this basis, a downscaling prediction model for extreme events in the Yangtze River Delta region, combining physical models and artificial intelligence, will be developed. This model will fully utilize global climate prediction data provided by the Intergovernmental Panel on Climate Change (IPCC).

5.1.3 *Monitoring and prediction of sea level changes in the Yangtze River Delta region*

The quantification of the magnitude and impact of sea level rise, along with the implementation of risk prevention measures, poses global challenges in addressing climate change and sea level rise. By leveraging real-time data and remote sensing technology, we continuously monitor and document changes in both global and regional ocean water levels. The online monitoring system for sea level changes is designed to provide timely, accurate, real-time information on sea level height, enabling the early identification and response to fluctuations, rises, falls, and changing trends in sea level. Real-time data on ocean water levels can be obtained through satellites, buoys, submersibles, radar, and other sensing devices. Satellite remote sensing technology directly measures sea surface height, providing global information on sea level and its changes with high spatiotemporal resolution.

Online monitoring of sea surface changes facilitates the acquisition of real-time trends in sea level variations, enabling an understanding of the speed and patterns of sea level rise or fall. This is crucial for monitoring global climate change, protecting marine ecosystems, and managing risks in coastal communities. The online prediction of sea level changes utilizes real-time monitoring data and advanced technologies, such as numerical models and machine learning algorithms, to model and predict future sea level changes. Numerical models simulate ocean processes, such as ocean currents and sea surface temperatures, providing a theoretical basis for predicting future sea level changes. Machine learning algorithms enhance the accuracy of these predictions by learning patterns from historical data.

The online monitoring and prediction system for sea surface changes offers timely and accurate sea level information to scientists, governments, businesses, and the public. It aids in the early detection of sea level rise, storm surges, and other oceanic events, providing early warnings and crucial support for response measures to address the impact of climate change on coastal areas. Suggestions for enhancing this project include the following:

(1) **Developing fusion processing technology for multi-source observation data**: Combining multiple monitoring technologies, including remote sensing, satellite altimetry, and radar altimetry, ensures comprehensive and multi-level data collection. Additionally, developing multi-source data fusion, processing, and analysis technologies will improve monitoring accuracy and spatiotemporal resolution.
(2) **Strengthening the ground station network**: While online monitoring relies on satellite and remote sensing technology, ground stations are also pivotal. Strengthening the network of groundwater level monitoring stations and improving the

verification and calibration of real-time monitoring data will ensure the reliability of the monitoring system.

(3) **Establishing a global cooperation network**: As sea level change is a global issue, it is recommended to establish a global cooperation network to share monitoring data and technical experience. This will aid in better understanding global sea level dynamics and improving the overall level of monitoring and prediction.

(4) **Developing flexible prediction models**: In the prediction phase, it is advisable to utilize flexible and adaptable prediction models, such as machine learning algorithms, which can learn from constantly evolving monitoring data and promptly adjust prediction results.

(5) **Establishing a real-time warning system**: By integrating online monitoring and prediction results, a real-time sea level change warning system should be established to promptly notify governments, communities, and the public in case of emergencies, enabling the implementation of necessary emergency measures.

5.1.4 *Chinese space reference antenna*

Given the significance of global, regional, and local monitoring of sea level changes, a robust spatiotemporal reference is essential. Therefore, it is proposed to construct infrastructure, namely the China Space Reference Antenna, consisting of a constellation of several small satellites in medium and low Earth orbits. The objective is to address the accuracy bottleneck faced by the technical performance and scientific application of the spatio-temporal reference, which is currently constrained by surface disturbances. The primary scientific objectives encompass establishing a spatiotemporal reference with the highest international accuracy of 1 mm and 0.1 mm/year; conducting gravitational redshift verification experiments based on star-ground and inter-star data using high-precision spatiotemporal benchmarks; advancing the level of relativistic verification by orders of magnitude; and enhancing our understanding of spatiotemporal geometry.

(1) **Domestic and international development trends**: Currently, the establishment of international spatio-temporal benchmarks relies

on ground observations, and the capacity to correct various gravitational and deformation effects of the Earth has nearly reached its limit, making further improvements in accuracy challenging. Consequently, the highest accuracy of current spatio-temporal benchmarks is approximately nanosecond (time) and centimeter (space) levels. The integration of multiple technologies and the establishment of spatial-temporal benchmarks will fundamentally overcome the limitations of existing observations, leading to improved indicators and broader scientific applications.

(2) **The role in future national strategy, technology, and social development:**

(i) *Promoting the development and advancement of scientific equipment R&D technology*: This program will be the first satellite in this field globally, carrying miniaturized and ultra-high-precision on-board very long baseline (VLBI)/satellite navigation (GNSS)/laser ranging (SLR) equipment, as well as precision measurement equipment such as electrostatic levitation accelerometers and optical clocks. The implementation of this mission will drive the development of related R&D technologies.

(ii) *Leading international scientific research in related fields*: By integrating multiple observation technologies, the space-based spatiotemporal reference scientific satellite will address issues caused by ground gravity, deformation, etc., unprecedentedly improving the accuracy of the spatiotemporal reference and supporting a series of scientific experiments, thereby achieving international leadership in this field.

(3) **Existing preliminary work and status of the technology team**: At present, domestic Very Long Baseline Interferometry (VLBI), Global Navigation Satellite System (GNSS), Satellite Laser Ranging (SLR), electrostatic suspension accelerometers, optical clocks, and other equipment have been acquired and installed on scientific satellites. Related research and exploration are also ongoing in data processing and scientific applications.

(4) **National strategy and scientific and technological objectives**: Based on the C-GRASP satellite, we aim to establish a spatiotemporal reference space benchmark that achieves the highest international precision of 1 mm/year and 0.1 mm/year, providing

benchmarks of higher precision for national strategies such as satellite navigation and deep space exploration.

5.1.5 *Development of integrated network data transmission and equipment for air, land, and sea in the Yangtze River Delta and adjacent sea areas*

(1) **Requirement Analysis**

The Yangtze River Delta urban agglomeration, situated near the Yellow Sea and East China Sea, spans a total of 41 cities across four provinces and municipalities: Shanghai, Jiangsu, Zhejiang, and Anhui. The region covers an area of 358,000 km^2 and has a population of 237 million. By 2023, the GDP of the Yangtze River Delta region will reach a staggering 30.51 trillion yuan, accounting for 24% of the national GDP and solidifying its position as a world-class urban agglomeration.

The area is well-connected by rivers, lakes, and seas, with abundant rainfall. However, the Yangtze River Delta is low-lying and flat, and factors such as transit floods, regional rainstorms, storm surges, and saltwater intrusion frequently overlap, leading to flood disasters and difficulties in water sourcing. Additionally, the Yangtze River Delta region boasts a developed economy, a high population density and industrial concentration, and key river mouths, such as the Yangtze River estuary, serve as the primary outlets for pollution in the basin.

To achieve disaster prevention and mitigation, water supply security, and water source protection in the Yangtze River Delta and adjacent sea areas, it is imperative to focus on the smart and high-quality water network of the Yangtze River Delta. This involves leveraging the integrated network of air, land, and sea within the region and adjacent sea areas to build a comprehensive water monitoring and perception network, as well as constructing the information infrastructure for the Yangtze River Delta water network.

(2) **Current Situation**

4G/5G Network: Currently, the integrated network of air, land, and sea in the Yangtze River Delta and adjacent sea areas is largely absent, lacking integrated planning. Some cities or regions supplement their water networks and points or surfaces based on ground communication networks like 4G/5G.

In 2021, Jiangsu Mobile initiated the "River, Lake, and Sea 5G Coverage Optimization" project, constructing and opening 307 5G base stations along the Jiangsu section of the Yangtze River. Through 5G multi-band collaborative networking, the coverage radius of 5G base stations on the river surface can exceed 6 km. Consequently, the Jiangsu section of the Yangtze River enjoys full 5G coverage, catering to the communication needs of the shipping industry, fisheries, tourism, emergency response, law enforcement, and other sectors.

The Shanghai Maritime Safety Administration has formulated and issued the "5G Network Offshore Coverage and Fusion Application" Action Plan (2023–2024), aiming to empower industries such as navigation assistance, wind power, and marine engineering facilities. The newly constructed 5G/4G base station at Jigujiao Lighthouse utilizes the high-quality 700 MHz frequency band, covering a radius of 25 km.

In January 2024, Hubei Province launched a 7.9 billion yuan infrastructure construction project for the Yangtze River Basin Information Expressway, encompassing the Yangtze River's one main stream and eight tributaries, as well as the shoreline of the Danjiangkou Reservoir. The project involves building 12,798 communication towers and utilizing tower facilities to mount various monitoring, surveillance, and sensing devices. Ultimately, it aims to achieve full 5G information coverage in key water areas of the Yangtze River Basin, establish a three-dimensional monitoring and perception system for the Hubei section of the Yangtze River Basin's water system, and preliminarily set up a control application system and monitoring collection system.

Satellite Constellation: The Eastern Insight Constellation is a joint venture between Wuhan University and the People's Government of Yantai City. The project plans to establish a global service system from 2027 to 2030, with an estimated 200 satellites in orbit. It aims to build a leading constellation operation system in China, encompassing "one satellite for multiple uses, multi-satellite networking, multi-network integration, and intelligent services," and achieve efficient information services for remote sensing data from acquisition to application on a global scale. The first experimental verification satellite of the Oriental Wisdom Eye constellation, Luojia 3-01, was successfully launched on January 15, 2023.

Furthermore, China is currently constructing the State Grid and G60 low-orbit giant constellation, with an expected completion of a network of 10,000 low-Earth-orbit satellites by 2030. Unlike Beidou's medium-to-high-orbit satellites, the giant low-orbit satellite constellation not only covers the globe but also offers high-speed data services and on-board computing processing capabilities. In terms of international standards development, the 3GPP Non-Terrestrial Network (NTN) has proposed two typical modes based on on-board payload capabilities: transparent forwarding and on-board regeneration. The development of transparent forwarding standards has been completed, and on-board regeneration modes will be further developed and enhanced in the future.

(3) **Problem Scenario**

The Yangtze River Delta encompasses a water system area of 23,937.2 km^2, including mainland coastlines of Shanghai, Jiangsu, and Zhejiang measuring 213 km, 954 km, and 2,253 km, respectively. Currently, the monitoring and perception of water systems and nearshore areas primarily rely on 4G/5G ground communication networks. However, the following issues persist:

Although the Jiangsu section of the Yangtze River and the Taihu Lake section have achieved full 4G/5G coverage, other areas remain under point coverage, resulting in a significant number of blind spots.

Extreme weather conditions, such as floods and typhoons, have resulted in 4G/5G disconnections. Due to water ingress, power outages, and fiber optic disconnections, the base station has been taken out of service. During the heavy rainfall in Changzhou in 2016, over 100 base stations were out of service due to water ingress and power failures. In 2021, 61,900 base stations were restored after being affected by the rainstorm disaster in Henan Province. In 2023, the Beijing–Tianjin–Hebei region was hit by a rainstorm, causing 1,600 base stations of Beijing Mobile alone to go offline.

Due to blind spots in wireless communication network coverage or the low communication rates of high-orbit satellites, offshore monitoring and maintenance remain manual processes, exhibiting very poor real-time performance.

(4) Suggestion

To build an integrated water network monitoring and perception network and information infrastructure for the Yangtze River Delta and adjacent sea areas, it is recommended to propose a construction plan for an integrated air, space, earth, and sea network in the region. Additionally, a prototype terminal for low-orbit satellites that integrates sensing capabilities must be developed. Furthermore, a satellite-based perception data implementation plan should be proposed for the Yangtze River Delta and adjacent sea areas, establishing transmission standards for perception data within that region.

5.1.6 *Development of a risk perception system for bank collapse in major artificial river mouths and channels in the Yangtze River Delta region*

5.1.6.1 *Evolution of main artificial river channels in the Yangtze River Delta*

An artificial river refers to a water conservation project excavated by human labor to connect water systems and facilitate transportation. It plays a crucial role in promoting social and economic development. In addition to natural rivers, the Yangtze River Delta region is crisscrossed by artificial rivers, making it the area with the highest density of river networks in the country. Over the past century, with the construction of numerous water conservation and waterway regulation projects in the Yangtze River Basin, extensive research has documented their impact on the geomorphological evolution of the lower reaches of the Yangtze River. However, there are limited research reports on their impact on the geomorphological evolution of the confluence of artificial rivers and mainstreams. The geomorphological evolution in this area has significant implications for surrounding shipping and coastal economic and social development and holds considerable practical and engineering significance.

The primary functions of artificial rivers include inland waterway transportation, flood prevention and control, water diversion and scheduling, and water landscape tourism. These artificial rivers in the Yangtze River Delta region crisscross like natural rivers, contributing to the region's high density of river networks. Typical examples include the Yangtze-to-Huaihe River, Yangtze-to-Chaohu River, Beijing–Hangzhou Grand Canal, Yangtze-to-Taihu River, the East Route of the South-to-North Water Diversion Project, Xuhe River, Huangpu River, and the Chongming Island Diversion Canal. (Ye Zhou, 2014; Liu Xiaoyan *et al.*, 2010; Qian Bijun, 2018). Due to their intersection with mainstream interests and engineering measures such as water diversion and scheduling, these artificial rivers face issues such as river channel evolution, channel siltation, slope damage, overloading of ship locks, and even saltwater intrusion from the Yangtze River estuary area.

The lower reaches of the Yangtze River generally feature bifurcated river channels (Yao Shiming *et al.*, 2003), numerous bends and river islands, and the evolution of river conditions has been dramatic throughout history (Pan Qingsang, 2010). The Yangtze River estuary, as a transitional zone for rivers entering the sea, is a coastal area lacking a distinct coastline and represents the most typical area of land-sea interaction (Chen Jiyu, 2000). Numerous islands and sandbars exist in the Yangtze River estuary, and the hydrodynamic conditions are complex and varied due to the combined effects of upstream runoff and offshore currents, leading to frequent geological evolution (Chen Jiyu *et al.*, 2008; He Yufang *et al.*, 2011). Since the last century, human activities have increasingly disrupted river systems. Large-scale water conservation projects, led by the Three Gorges Dam, and soil and water conservation measures have significantly reduced sediment transport in the lower reaches of the Yangtze River, slowing down the siltation and the expansion of the underwater delta at the estuary and causing local erosion and retreat (Li Peng *et al.*, 2007; Du Jinglong *et al.*, 2012; Yang *et al.*, 2015; Liu Jie *et al.*, 2017; Liao Chunyan, 2010; Wang Lihua and Zhou Yunxuan, 2018; Guo *et al.*, 2019). Furthermore, numerous waterway regulations, beach and bank protection measures, and water conservancy projects have been carried out in the downstream and estuary areas of the Yangtze River, altering the flow and boundary conditions of local river sections (Luan Hualong *et al.*, 2018; Liu Xiaobin *et al.*, 2011; Wu Shuaihu *et al.*, 2016; Zhang Xiaohe *et al.*, 2015). Frequent human remediation and development will inevitably have a significant impact on the geomorphological evolution of rivers and the ecological balance of watersheds (Chen Jiyu *et al.*, 2002).

Following the construction of a series of water conservation and remediation projects in the Yangtze River Basin, abundant research results have emerged regarding their impact on the geomorphological evolution of the lower reaches of the Yangtze River. However, there is a scarcity of research reports addressing their influence on the geomorphological evolution of the confluence areas between artificial rivers and main streams. The geomorphic evolution of these confluence sections significantly impacts navigation and coastal economic and social development around the outlets of artificial rivers. Furthermore, these patterns vary across different mainstream sections, making them a crucial aspect of the comprehensive management of the Yangtze River Basin. Therefore, exploring the geomorphic evolution characteristics and trends, particularly the evolution risks, of the main artificial rivers and their confluence with main streams in the Yangtze River Delta region is beneficial for flood control, waterway regulation, and comprehensive development and utilization within the Yangtze River Basin. It also aids in the comprehensive management and development of artificial rivers, possessing significant practical and engineering importance.

5.1.6.2 *Study on the evolution of river channel geomorphology*

(1) **Research methods for the evolution of river channel topography**: The evolution of river channel topography is a pivotal river surface process, primarily influenced by the interplay of water flow, sediment, and riverbed materials (Shen Yuchang and Gong Guoyuan, 1986). Traditional research on river channel landform evolution primarily relied on semi-quantitative analysis of measured data, such as water and sediment conditions (Yang Zhongyong, 2014). This involved analyzing the dynamic conditions and sediment transport characteristics of river channels from the perspectives of dynamic sedimentation and landform dynamics to elucidate the evolutionary characteristics of river channel landforms. In recent years, with the rapid advancement of computer technology, GIS numerical and analytical simulation methods have become widely utilized in this field.

(2) **Research on the evolution of river channel geomorphology based on actual measurement data**: Chen Jiyu *et al.* (1957, 1959) examined the development process of the Yangtze River Delta

from geological, sedimentary, and topographical perspectives, laying the groundwork for further comprehensive research on the Yangtze River estuary. Shen Huanting and Pan Ding'an (1979) detailed the nature, movement form, duration, flow velocity, flow field distribution, and the relationship between tidal currents and tidal levels in the Yangtze River estuary based on extensive measured data. They explored the interplay between tidal current characteristics and channel geomorphological evolution. Huang Sheng (1986) analyzed water flow and sediment characteristics in the Yangtze River estuary using measured data and concluded that the river channel's geomorphological evolution is marked by regular branching of sandbars and islands, south-north silt island migration, ebb tide main tide oscillation, and alternating port rise and fall in the north and south. Li Jiufa *et al.* (2006) analyzed sediment movement characteristics at the Yangtze River mouth and investigated the formation process and evolution mechanism of the nine-segment sediment. Wang Yonghong *et al.* (2009) collected sediment and water samples from the riverbed surfaces of the Xinqiao Channel, Nanxiaohong Creek, South Branch, and South Channel within the Yangtze River estuary tidal trough. They analyzed the sediment and dynamic characteristics of this tidal trough, laying the foundation for revealing its evolutionary characteristics. Chen Jun *et al.* (2010) examined the morphological characteristics, erosion and deposition types, and evolutionary trends of tidal flat profiles along the Jiangsu Province coast using measured elevation data and historical topographic data.

Moreover, as instruments and equipment continue to advance, field exploration methods have become increasingly diverse, and observation accuracy has seen significant improvement. Cheng Heqin *et al.* (2002, 2003, 2004) Li Weihua *et al.* (2008), Wu Shuaihu *et al.* (2016) Wu *et al.* (2016) Zheng Shuwei *et al.* (2016, 2018) Zheng *et al.* (2018) Shi Shengyu *et al.* (2017) Shi *et al.* (2018) and Wang Shuping *et al.* have contributed to this field through their research efforts.

(3) **Research on the evolution of river channel landforms based on GIS technology**: With the rapid advancement of computers and information technology, geographic information systems (GIS) and digital elevation models (DEMs) are increasingly being utilized in the study of river channel geomorphological evolution. Huang Jiazhu (1999) employed a multi-temporal remote sensing

dynamic analysis method to investigate changes in the riverbanks below Nanjing over the past 40 years and utilized GIS technology to study the evolutionary patterns of the river channel. Wu Hualin *et al.* (2002) developed digital elevation models of the Yangtze River estuary at various times using GIS, based on nautical chart data spanning from 1842 to 1997. They analyzed the evolution of the sandbank channel, shoreline alterations, and sandbar changes in the Yangtze River estuary by integrating the characteristics of longitudinal and transverse profile variations. Li Bochang (2006) examined changes in river morphology, longitudinal and transverse adjustments of the riverbed, and erosion and deposition patterns in the northern branch of the Yangtze River, utilizing digital terrain modeling technology based on the underwater topographic map of the Yangtze River estuary. Li Fang (2008) integrated GIS technology with sediment transport mechanics, engineering hydrology, and coastal dynamics to quantify the erosion and deposition evolution of the Yangtze River estuary before and after the construction of the Three Gorges Project, and analyzed its underlying causes. Liu Dujuan *et al.* (2010) digitized six nautical charts from 1980 to 2005 using GIS technology, established a digital elevation model, and assessed the evolution and stability of the river morphology in the southern branch of the Yangtze River estuary. Wu Shuaihu (2017) created a digital elevation model using GIS technology, based on extensive historical nautical chart data, and conducted a comparative analysis of the river channel evolution patterns of the Yangtze River estuary and the Mississippi River estuary, as well as their response to human activities. Wang Shuping *et al.* (2018) digitized underwater terrain data from 1998 and 2013 using GIS, established a DEM and combined it with multibeam depth measurement data to explore the erosion and deposition evolution and micro-geomorphic characteristics of the Yangtze River and Poyang Lake confluence. Sun Yi (2018) identified the characteristics and driving factors of erosion and deposition in the Yangtze River estuary over the past 20 years using historical and measured water depth data, GIS methods, and factors such as basin inflow, sediment, engineering construction, and sea level rise. The integration of GIS and DEM technologies enables both qualitative and quantitative analysis of river channel erosion and deposition changes, providing a highly visual representation of the evolutionary process. This has emerged as a relatively mature and reliable approach to the study of river channel geomorphological evolution.

(4) **Research on the geomorphic evolution of the lower yangtze river channel**: Given that the longitudinal profile of the lower reaches of the Yangtze River is stepped (Institute of Geography, Chinese Academy of Sciences, 1985), falling into the category of a relatively stable bifurcated channel (Pritchard, 1967), changes in plan shape better reflect the evolutionary characteristics of channel geomorphology (Seminara *et al.*, 2001; Luchi, 2010). The adaptive adjustments of the channel result in significant longitudinal differences in erosion and siltation patterns and intensities. Lin Linsong *et al.* (2006), Lv Lijun *et al.* (2009), Liu Xiaobin *et al.*, and Wang Xiangdong *et al.* (2008) established a geographic information system for the Zhenyang section of the lower reaches of the Yangtze River within a GIS environment and studied the erosion and deposition evolution process in this section through a river channel erosion and deposition analysis system. Li Maotian (2005) calculated the erosion and deposition in the lower reaches of the Yangtze River, revealing notable differences in the evolution of river channel topography along its course. Zuo Liqin and Lu Yongjun (2014) analyzed the controlling role of nodes in the evolution of the channel within the Madang section of the lower reaches of the Yangtze River through their research. Yuan Xiaoting *et al.* (2019) investigated the characteristics of tidal dynamics changes in the lower reaches of the Yangtze River from Datong to Nanjing and found that recent overall channel deepening in the Yangtze River's tidal section has led to a decrease in upstream tidal wave resistance.

As the frequency of human activities increases, the geomorphic evolution of the lower reaches of the Yangtze River has garnered greater attention. Qu Guixian (2014) explored the channel evolution characteristics and response patterns of the Datong to Jiangyin section of the Yangtze River, which is sensitive to changes in the tidal boundary and tidal zone boundary before and after the completion of the Three Gorges Project. Zheng Shuwei *et al.* (2018) analyzed the evolutionary characteristics of the underwater terrain from Datong to Xuliujing in the lower reaches of the Yangtze River using historical and measured data. The construction of the Three Gorges Dam resulted in a sharp decrease in upstream sediment, leading to overall erosion of the river section. Additionally, shoreline utilization and channel improvement projects narrowed the river channel, intensifying bank slope erosion. Xu Wei *et al.* (2019) pointed out that, under the influence of human activities, such as the Three Gorges Dam, the Nanjing section of the lower reaches of the Yangtze River has exhibited overall scouring characteristics over the past 20 years. They analyzed micro-geomorphic features, including sand waves, underwater steep slopes, and bridge pier scouring pits, as well as the future evolutionary trend of the channel.

(5) **Study on the geomorphic evolution of the Yangtze river estuary channel**: Previous researchers have conducted extensive studies on the geomorphological evolution of the Yangtze River estuary. Chen Jiyu *et al.* (1979) summarized the development pattern of the Yangtze River estuary over the past two thousand years into five key aspects: siltation and expansion of the southern bank beach, merging of sand islands on the northern bank, narrowing of the estuary, formation of the river channel, and deepening of the riverbed. Since the end of the last century, a series of large-scale projects constructed in river basins and estuaries have responded to the geomorphic evolution of estuarine areas (Yang *et al.*, 2011; Luan *et al.*, 2016). The entire Yangtze River estuary continues to undergo adaptive adjustments to the new water and sediment environment, gradually achieving a new equilibrium (Zhang Xiaohe *et al.*, 2015), resulting in an underwater beach erosion pattern (Ji Na *et al.*, 2013; Li Bing *et al.*, 2015). Erosion and volume expansion in the upper sections of the southern branch, southern port, and northern port of the mouth have directed channel morphology toward a relatively narrow and deeper configuration (Liu Jie *et al.*, 2017).

The implementation of the North Passage Deep Waterway Project has involved the erosion of the upper section of Nanhui Beach, siltation of the lower section, overall net erosion of the North Channel, and siltation of the lower Jiuduansha Shoal along the embankment (Liu Jie *et al.*, 2010; Pan Lingzhi *et al.*, 2011; Fang Qiangfei *et al.*, 2014).

The influence of the North-South Channel diversion control project, primarily centered on the Qingcaosha Reservoir and large-scale projects such as the Shanghai Yangtze River Bridge, on the geomorphic evolution of the Yangtze River estuary channel is also significant (Zheng Wenyan *et al.*, 2010; Zhou Falin, 2012; Sun Yi *et al.*, 2018). Following the implementation of the North-South Channel diversion project, a slightly curved and smooth channel, named "South Branch Main Channel—New Bridge Channel—North Channel" was formed, enhancing the hydrodynamic conditions of the diverted North Channel (Wu Yan *et al.*, 2016). The shoal of the Central Sand—Xin Liuhe Sand remained essentially stable, and the trend of upper shoal's erosion and retreat was halted (Ruan Wei *et al.*, 2011). The downstream river conditions of the diversion port remained relatively stable (Song Shaohong *et al.*, 2007; Lu Zhongmin *et al.*, 2009). The construction of the Qingcaosha Reservoir narrowed the upper and middle reaches of the North Channel, causing deep water to flow northward and severely eroding the Liuyao Sand Ridge on the south bank of Chongming Island, thereby enhancing the ebb tide's influence in Beigang (Liu Weiyi *et al.*, 2011; Liu Gaowei *et al.*, 2015; Guo Xingjie *et al.*, 2016). Additionally, the construction of the Shanghai Yangtze River Bridge narrowed the river channel, leading to significant erosion of the middle and upper sections of the North Channel (Mo Ruoyu *et al.*, 2015), and a "double kidney-shaped" local erosion landform was observed near the main pier of the bridge (Lu Xuejun *et al.*, 2016).

5.1.6.3 *Confluence section of artificial and natural rivers*

The confluence section of an artificial river and a natural river refers to the area where the artificial river merges with and intersects the main stream of the natural river. Li Rui *et al.* (2005) conducted a two-dimensional, single-frequency wave field numerical simulation study on Wusongkou using the CGWAVE steady-state wave gentle slope equation model. Zhou Yicen *et al.* (2019) analyzed the reasonable estimation values of the low tide level recurrence period at Wusongkou using mathematical statistics and time-varying moment methods, based on the lowest tide level sequence from 1978 to 2015 at the Wusongkou tidal station. Song Zhaojun *et al.* (2008) interpreted historical remote sensing image data near Jiaozhou Bay and Laizhou Bay, analyzed the geomorphic evolution characteristics of the southern and northern entrances of the Jiaolai Canal, and explored the potential environmental impacts following the opening of the Jiaolai Artificial Canal. Dou Xuesong and Zhu Cunzhang (2010) proposed methods and key points for quality management of the dredging project at the Liuwei River mouth of the Beijing–Hangzhou Grand Canal, based on years of experience in measurement and dredging engineering implementation, and summarized comprehensive operating procedures and technical specifications. Yu Feng (2016) used a combination of two-dimensional mathematical models and fixed-bed physical models to simulate the water flow characteristics at the intersection of the North Canal and Liangshui River in Beijing. Overall, there is relatively limited research on the confluence of artificial and natural rivers, particularly regarding the geomorphic evolution characteristics of typical artificial river and main stream confluence sections in large river basins.

5.1.6.4 *Research on underwater microgeomorphology*

Underwater micro-topography refers to small-scale features commonly developed on the bed surface of rivers, estuaries, and shallow sea areas (Allen, 1982; Ashley, 1990; Cheng Heqin *et al.*, 2003; Allison *et al.*, 2012; Salvatierra *et al.*, 2015). Wu Shuaihu *et al.* (2016) suggested that its formation may be related to the interaction between near-bottom flow and irregular bed topography (Simarro *et al.*, 2015); it exhibits characteristics of short evolutionary time and high frequency. The morphological evolution of underwater micro-topography can be assessed from a micro perspective to evaluate riverbed stability (Cao Lihua *et al.*, 2006; Wang Weiwei *et al.*, 2007), further revealing the shaping process of macro-topography (Zhao Baocheng, 2011). It largely reflects changes in hydrodynamic fields and sediment conditions (Wu *et al.*, 2009) and is a commonly used indicator for studying dynamic geomorphic processes in watersheds, estuaries, and coastal areas (Shi Shengyu *et al.*, 2017). The shaping and distribution of underwater micro-landforms are primarily controlled by sediment transport, a theory widely applied in fluid geomorphology research (Allen, 1982; Best, 2005).

There are numerous types of underwater microgeomorphology, with sand waves being a focal point of research for scholars globally. Sand waves are rhythmic microgeomorphology primarily formed in sedimentary environments ranging from medium sand to gravel (Kostaschuk, 2000). Their morphological resistance can diminish the sediment transport capacity of water flow, and their formation and evolution can indicate corresponding changes in dynamic regional conditions (Du Xiaoqin *et al.*, 2008).

Foreign experts have conducted extensive research on sand wave in various large river basins. Allen (1980) systematically explained the internal structure of sand waves and their associated surface water flow patterns, noting that the internal structure depends on the asymmetric strength of water flow and tidal currents. The asymmetrical shape of sand waves is attributed to the asymmetrical fluctuation of tidal currents. Soulsby *et al.* (1991) studied sand waves in the Torric River's mouth and discovered no flow separation at high velocities but intermittent flow separation at lower velocities. Kostaschuk and Villard (1996) found that the peak shape of sand waves in the Fraser River correlates with the sediment transport rate. Parsons *et al.* (2005) examined sand waves in the Paraná River in Argentina and observed that sand waves with saddle-shaped ridges exhibit larger and structurally stronger vertical velocity separation regions compared to those with straight-shaped ridges. Damen (2018) emphasized that sand wave development is influenced not only by riverbed morphology and hydrodynamics but also by suspended sediment transport, which can inhibit sand wave height growth.

Research on sand waves in China began relatively later but has already yielded significant results. Ren Mingda and Wang Nailiang (1985) attributed the main causes of sand waves to wind, current, and wave action. Cheng Heqin *et al.* (2000, 2002, 2003, 2004) used high-resolution underwater detection instruments to observe the Yangtze River estuary's main channel, analyzing sediment transport characteristics and micro-topography of sand waves. They found that small-scale sand waves form when the flow velocity is 0.6–1.0 m/s, and large-scale sand waves often form when the flow velocity exceeds 1.1 m/s. They also proposed a new boundary for the transition between sand ripples and sand waves, linking sand wave morphology and sediment particle size on the bed surface of the southern branch of the Yangtze River estuary's South Channel to seasonal runoff. Zhong Liang *et al.* (2013) conducted a fixed-bed experimental study on sand wave resistance from a fractal perspective. Zheng Shuwei *et al.* (2016) discovered "chain bead-shaped sand waves" near the North-South Port diversion at the Yangtze River's mouth, composed of chain-shaped sand waves and elliptical depressions, and preliminarily explored their formation reasons.

In addition to sand waves, studies on other underwater micro-morphology exist. Chen Weimin *et al.* (1992) identified various micro-morphology on the underwater bank slope of the Yellow River estuary, including flatbeds, pits, depressions, gullies, residual sand dunes, and disturbed composite bottom beds. They also found flat beds and giant sand waves on the underwater bottom slope of the Yangtze River estuary, along with micro-morphology such as air pockets, erosion marks, and sediment patches. Zhao Baocheng *et al.* (2011) discovered four types of micro-morphology on the underwater bank slope of Hangzhou Bay's north bank: smooth bottom beds, gullies, pits, and sand waves. Shi Shengyu (2017) conducted navigational observations on recent changes in the Yangtze River's tidal boundary and found erosion micro-morphology such as scour pits, scour deep troughs, underwater erosion bank slopes, and composite erosion bank slopes. Zheng Shuwei *et al.* (2018) studied erosion landforms of typical bridge piers in the middle and lower reaches of the Yangtze River using a multibeam depth measurement system. Wang Shuping (2018) examined the confluence of the Yangtze River and the Dongting and Poyang Lakes, finding that the bed surface of the confluence has developed micro-morphology such as flat beds, sand waves, erosion marks, erosion pits, and grooves in recent times. The response of micro-morphology development to riverbed evolution was also explored. Huakai *et al.* (2019) conducted field investigations and found large nearshore scour pits in the Yangtze River estuary's horizontal sand channel, exploring their formation mechanism in conjunction with engineering construction.

5.1.6.5 *Intelligent identification of nearshore disaster geomorphic risks*

Disaster geomorphology refers to geological processes and phenomena that pose a threat to human lives and production activities (Liu Xilin and Zhao Yuan, 2008). Bank collapses are a prevalent type of disaster landform along major rivers globally, posing a risk to embankment safety, causing land loss along riverbanks, and potentially disrupting river stability over the long term, leading to irreparable consequences. A bank collapse can significantly contribute to the erosion and retreat of shorelines and shallow waters. It manifests as a result of riverbed evolution, specifically the erosion or collapse of the floodplain within the river channel due to the interplay of water flow and riverbed processes (Zhang Xingnong *et al.*, 2008).

Bank collapses narrow the width of the shallow water area adjacent to the embankment facing the river channel, prompting the main stream to directly impact the embankment's foundation through top-flushing or localized scouring. Additionally, bank collapses frequently damage the protective layer of bank slopes, leading to potential embankment foundation leakage and increasing the risk of embankment failure during rapid water level fluctuations during major floods (Li Chang'an *et al.*, 2008).

Bank collapses pose a significant challenge in the middle and lower reaches of the Yangtze River, garnering considerable attention due to its severe threat to the safety of the Yangtze River's flood control embankments and the stability of the river system. For instance, in November 1976, a landslide measuring 460 m in length and 350 m in width occurred before the Ma'anshan Power Plant, resulting in a 450 m collapse of the river embankment and seriously jeopardizing the power plant's safety (Li Chang'an *et al.*, 2008). By the 1990s, the collapsed section of the Yangtze River extended over 1,520 km, accounting for 35.7% of all riverbanks in the middle and lower reaches (Yangtze River Water Resources Commission, 1990).

Following the operation of the Three Gorges Reservoir in 2003, incidences of bank collapse remained severe. According to preliminary statistics from the Yangtze River Water Resources Commission, between 2003 and 2012, a total of 655 riverbanks collapsed along the middle and lower reaches of the Yangtze River, spanning a combined length of 495.9 km (Ministry of Water Resources of the People's Republic of China, 2011). In recent years, sudden shoreline collapses continue to occur and cause significant harm. For example, on November 8, 2017, a major landslide struck Zhinan Village in Yangzhong, breaching the Yangtze River flood control embankment and damaging several residential houses (Figure 5.1). Large-scale landslides, extending further inland, can breach embankments.

Jiangsu Province rapidly collapsed. Within a short period, houses, villages, docks, fields, and warehouses near the shore were either

Figure 5.1. An embankment along the Yangtze River in Yangzhong City on November 8, 2017.

dragged into the river or caused the river water to surge back and engulf residential areas, resulting in significant losses and severely impacting land planning, utilization, and the daily lives of residents. With the adoption of multi-beam underwater detection technology, the influence of underwater steep slopes, scour holes, and sand waves on riverbank stability has garnered increasing attention. When the local scouring depth of the riverbed is substantial, a steep slope forms around the scour hole. If the scour hole is situated close to the shore, it steepens the underwater bank slope, creating a precarious slope. Such steep riverbank slopes are susceptible to instability due to water flow and wave action, ultimately leading to complete bank slope collapse.

Erosion pits and steep slopes in the nearshore areas of the middle and lower reaches of the Yangtze River have developed across multiple sections, accompanied by bank collapse and posing a risk to floodplain stability. For instance, Shi Shengyu *et al.* (2017) utilized navigation maps and multi-beam data to study underwater erosion terrain from Datong to Wuhu and discovered that multiple erosion pits developed concurrently with steep underwater slopes. They suggested that the development of nearshore erosion pits and steep slopes could jeopardize the safety of embankments, revetments, and other engineering projects. Zhang Jiahao *et al.* (2018; 2019) measured the onshore and underwater topography of the Ma'anshan River floodplain slope and identified an elliptical scour pit approximately

3 m offshore, with dimensions of 69.2 m in length, 44.6 m in width, and 19.2 m in depth. The slope of the erosion pit's edge exceeds 40°, forming a steep ridge. Huakai *et al.* (2020) analyzed the geomorphological evolution of navigation maps and multi-beam measured micro-geomorphology at the north entrance of the Hengsha Channel in the Yangtze River estuary over multiple phases, finding that the erosion pit at the channel entrance has been deepening, widening, and approaching the east bank annually, creating high and steep slopes that threaten coastal stability.

Sand dunes are unstable bedforms that alter the water flow structure. Especially when sand waves develop near the offshore slope, they modify the slope's foot terrain, impact the flow field, drive changes in river patterns, and consequently affect riverbank stability. Furthermore, sand waves and their movements pose several risks. Firstly, the presence and migration of large sand waves can reduce local navigation depth, compromising navigation safety (Aliotta and Perillo, 1987; Knaapen and Hulscher, 2002). Frequent dredging may be necessary due to the decrease in minimum navigation depth caused by unstable sand waves that tend to reform after dredging (Knaapen *et al.*, 2001; Besio *et al.*, 2008). Secondly, the safety of underwater pipelines, communication cables, and offshore structures relies on seabed stability. Sand wave movement can damage submarine pipelines (Morelissend *et al.*, 2003) or block inlet valves, exposing and creating unsupported spans in submarine cables and oil and gas pipelines, leading to malfunctions. Large sand wave migrations can even cause tunnel exposure, threatening its structural integrity. Lastly, sand wave migration is also a risk factor for the stability of oil platforms and wind farms (Blondeaux and Vittori, 2016). The Yangtze River Economic Belt is one of China's regions with the strongest comprehensive strengths and strategic support (Jiang Yuehua *et al.*, 2017). The mainstream of the middle and lower reaches of the Yangtze River, a vital golden waterway connecting the east and west, links important cities such as Yueyang, Wuhan, Jiujiang, Nanjing, and Shanghai. As underwater engineering construction advances in the middle and lower reaches of the Yangtze River, the disaster risks posed by sand waves will increasingly come to the forefront. Therefore, this article collectively refers to underwater disaster landforms as bank collapses, steep slopes, erosion pits, and sand waves.

Determining the extent to which geomorphology evolves into disaster landforms presents a scientific challenge. Identifying underwater disaster landforms has become crucial for early warning of riverbank instability. However, advanced underwater terrain data acquisition technology, dominated by multi-beam depth measurement systems, has been relatively late in gaining widespread adoption. The interpretation of underwater disaster terrain information in high-resolution data still primarily relies on visual recognition, significantly hindering the efficiency and practical application of identifying underwater disaster micro-landforms.

5.1.6.6 *Development of a risk perception system for major artificial river mouth collapse*

In recent years, the frequency and intensity of artificial river mouth collapses have increased unexpectedly, primarily due to their triggering and disaster mechanisms being tied to factors such as sediment reduction, nearshore engineering, and severe weather conditions. There is an urgent need to transcend traditional research paradigms and enhance theoretical research on the evolution of nearshore disaster geomorphology through data-driven models. The technical hurdle lies in the significant lack of three-dimensional, fine terrain and sediment panoramic data in the intertidal zone, a limitation of current domestic and foreign sound and light detection instruments. Therefore, it is recommended to expedite the development of an intelligent perception system for severe disaster risks in estuarine and nearshore shoals, integrating sound and light linkage and sea-air integration. This system should address the challenges of collaborative detection by heterogeneous dual agents, enable 360° real-time acoustic imaging ranging, achieve high-precision detection of echo photons with transmission losses exceeding seven orders of magnitude, and incorporate optimization technology for multipath, non-line-of-sight wireless channels. It should be capable of obtaining panoramic data from shallow to extremely shallow waters at depths ranging from 0 to 50 m, revealing the geological mechanisms behind rapid erosion, bank collapses, scouring, and erosion troughs. Additionally, it should realize a sub-meter level of risk identification, facilitate online risk perception, and provide crucial technical and instrumental support for disaster risk prevention and control of nearshore infrastructure.

5.1.7 *Development of intelligent and high-quality water quality parameter over-limit detection equipment in the Yangtze River Delta*

5.1.7.1 *Monitoring and analysis of water quality in the Yangtze River Delta water network utilizing satellite remote sensing technology*

Satellite remote sensing represents a pivotal technical approach for monitoring water quality in coastal areas, rivers, and lakes within the Yangtze River Delta. However, there are consistency issues among products sourced from different satellites, particularly concerning the stability of satellite data for nearshore, lake, and river channels under varying surface conditions, as well as the accuracy of water quality monitoring data products for terrestrial rivers and lakes amid complex atmospheric conditions. Consequently, it is imperative to conduct authenticity testing of remote sensing products within the Shanghai region and undertake research on refined radiometric calibration of satellite data in specific areas. This will provide technical support for verifying the authenticity of satellite remote sensing monitoring of water quality in the nearshore and river-lake zones of the Yangtze River Delta. Authenticity testing should encompass diverse domestic and international remote sensing satellite water quality detection products to ensure the reliability of remote sensing monitoring services for water bodies in the region.

(1) Main Research Content

Investigate the model for satellite remote sensing-based real-time water quality monitoring in coastal areas, rivers, and lakes within the Yangtze River Delta, incorporating low-cost, low-power, and durable customized buoy technology, to conduct integrated application research on a ground-based water quality detection system.

Address the challenges associated with water quality sampling in monitored water bodies, soil erosion, and illegal pollutant discharges into rivers, lakes, and seas by researching water quality sampling methodologies. Construct a spectral library of water quality samples from the coastal areas, rivers, and lakes of the Yangtze River Delta, and establish a multi-frequency sampling verification system to achieve real-time and online water quality sampling and verification across the entire Yangtze River Delta coastal, river, and lake areas.

Research the minimum buoy deployment plan based on the hydrological characteristics of Shanghai's coastal and river networks. Select typical rivers and lakes to study the spatiotemporal diffusion model of polluted water bodies and overcome the challenge of spatiotemporal coverage in polluted water body diffusion sampling.

Develop an automated authenticity verification algorithm based on multi-source sampling data to address the issue of authenticity verification for large-scale water area, water quality, illegal discharge, and soil erosion monitoring spatial big data derived from satellite remote sensing data. Conduct research on the fine calibration of satellite radiation in the Shanghai area to address the accuracy of satellite data.

Based on the requirements of the river chief system and the commercialization of river and lake health assessments in Shanghai, research needs to be conducted on the integration of automated river and lake health monitoring application systems, accompanied by relevant application demonstrations.

(2) System Platform Construction

The system platform integrates 3S technology, mobile internet, the Internet of Things, cloud computing, and other technologies, utilizing the "one cloud, multiple terminals" management approach. Combined with automated monitoring equipment, it supervises the coastal and river-lake areas of the Yangtze River Delta, enabling real-time data transmission, business management processes, intelligent supervision and assessment, and automated warning and notification. This enhances response speed in addressing coastal and river-lake governance issues.

Main performance indicators of the system: Supports six data collection and access methods: satellite remote sensing data, satellite navigation data, aerial remote sensing data, mobile office terminal data collection, vehicle-mounted mobile measurement data, and traditional hydrological and water quality sensor data access.

Supports eight types of remote sensing monitoring products: Hydrogeological surveys, water resources evaluations, wetland resource distributions, assessments of eutrophication intensity, suspended solids monitoring, soil and water conservation remote sensing monitoring, soil and water engineering remote sensing monitoring, and flood disaster monitoring.

Supports two water big data analysis models: Water pollution diffusion prediction analysis and urban flood simulation analysis.

Supports deployment forms: WEB end and mobile end.

Provides real-time responses to various human–computer interaction operations, information queries, graphic operations, and other software system functions. The information query, operation, and input interface should be displayed on the computer in three ways: graphically, textually, and with data. The data table should have the function of generating reports. The system's operational requirements are designed to be simple and user-friendly.

Utilizes WebGIS to execute GIS analysis tasks: Accessing map services via standard browsers (such as Internet Explorer) necessitates visual processing and querying of river status and warning responses on GIS, along with seamless scaling capabilities.

(3) Monitoring and Analysis of Water Quality in the Yangtze River Delta Coastal Areas, Rivers, and Lakes Using Multispectral/ Hyperspectral Technology

Conduct on-site measurements and analyses of water spectra within the Yangtze River Delta region. Based on the collected water quality data and spectral measurements, align the measured remote sensing reflectance with corresponding bands for lightweight and small multispectral/hyperspectral sensor bands. Establish a statistical regression model and utilize remote sensing imagery to quantitatively invert the water quality of the Yangtze River Delta region. Investigate the fusion of multispectral/hyperspectral satellite image data with lightweight, small-scale multispectral/hyperspectral sensor data, analyze the large-scale distribution characteristics of water quality, and provide rapid decision-support information for water environment monitoring and water quality assessments in the Yangtze River Delta region. The primary focus of the research is as follows:

Research on Water Quality Data Acquisition Methods Using Lightweight and Small Multispectral/Hyperspectral Sensors: Investigate the method of acquiring water quality data using lightweight and small multispectral/hyperspectral sensors to address the challenge of rapid data acquisition in unstable environments. Conduct research on stable imaging techniques for these sensors in complex wind speed environments and complete corrections for errors such as data acquisition position, attitude, and eccentricity. Analyze the variations in multispectral/hyperspectral data within the micro-dynamic environment of river channels, study the stitching and mosaicking methods for multiple bands and images of water quality, and resolve the issue of mapping large-scale water quality multispectral/hyperspectral data in the Yangtze River estuary area.

Simultaneous On-site Observation of Existing Monitoring Sections along the River Channels: Collect hydrological characteristics of typical river channels in Shanghai and utilize existing river section monitoring data to conduct on-site, synchronous observations and collections of river and lake water column. Relevant instruments include dual-channel spectrometers, flow velocity and direction meters, solar photometers, wind speed and direction meters, and water sample filtration devices. Subsequent laboratory analysis parameters primarily include water remote sensing reflectance, flow velocity, flow direction, aerosol optical thickness, particle size, and mineral composition.

Quantitative Inversion of Water Quality Indicators Based on Lightweight and Small Multispectral/Hyperspectral Data: By performing correlation analyses between measured water quality indicators and remote sensing reflectance data, identify the bands that are most sensitive to water quality. Calculate remote sensing reflectance and establish a quantitative inversion empirical model for water quality indicators using statistical regression analysis methods. Utilize the established model to invert water quality from multispectral/hyperspectral data and analyze the inversion results.

Water Quality Index Inversion Based on Satellite Image Multispectral/Hyperspectral Data and Fusion of Lightweight Multispectral/Hyperspectral Sensor Data: For water quality monitoring in Shanghai, develop a water quality index inversion model using multi-source data fusion, including satellite data and data from the lightweight and small multispectral/hyperspectral sensors. Based on multi-period and multi-type multispectral/hyperspectral sensor data, determine the water quality status of the Shanghai waters and analyze the spatial and temporal distribution and changes in Shanghai water quality.

5.1.7.2 *Shanghai water affairs marine digital twin water network construction*

(1) Digital Twin Water Network

The digital twin water network is a process of digital mapping, intelligent simulation, and forward-looking prediction of the physical water network, grounded in the physical water network. It utilizes spatiotemporal data as the foundation, mathematical models as the core, and water conservation knowledge as the driving force (Figure 5.2).

(2) Digital Twin Platform: Data Acquisition (Figure 5.3).

(3) Internet of Things Perception System (Figure 5.4).

(4) Research and development of digital twin core technology system (Figure 5.5).

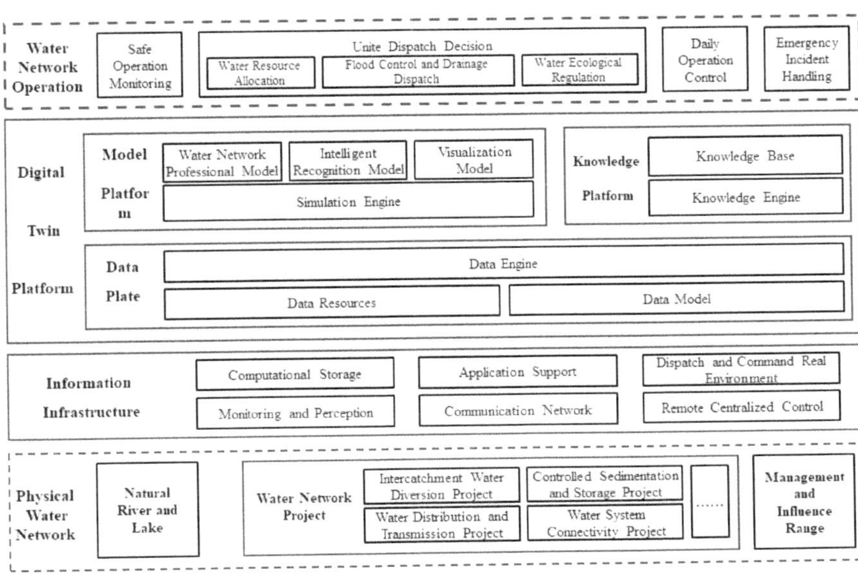

Figure 5.2. Framework diagram of digital twin water network technology.

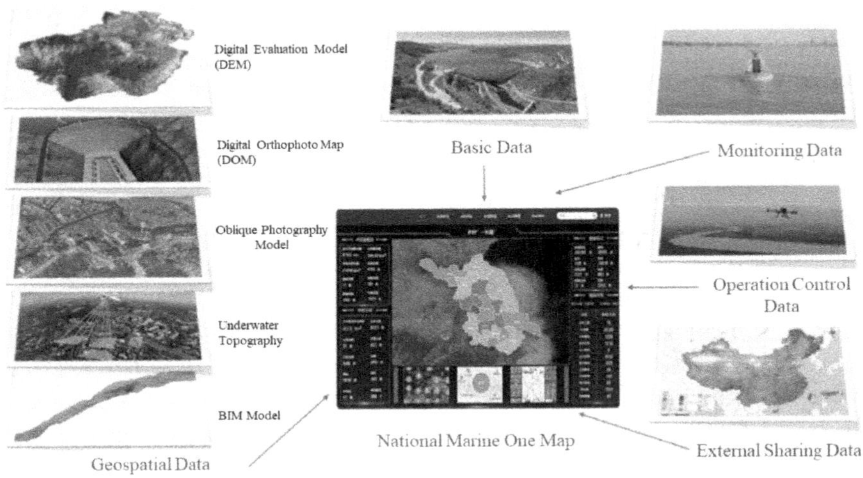

Figure 5.3. Digital twin platform: data acquisition diagram.

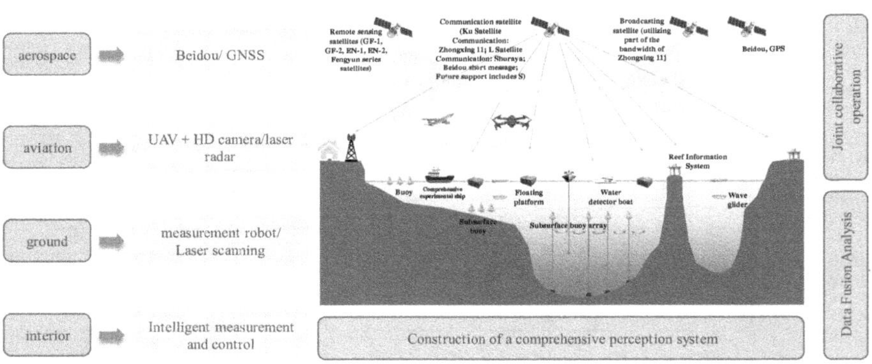

Figure 5.4. IoT perception system diagram.

5.1.7.3 *Portable fluorescent sensor array*

Initiate the development of over-limit detection equipment for crucial components such as organic matter, heavy metals, and algae within the smart and high-quality water network of the Yangtze River Delta. This involves creating novel water network element detection equipment designed to swiftly and precisely detect these components (including organic matter, heavy metals, algae, etc.) specifically for the purpose of intelligent and high-quality water network monitoring.

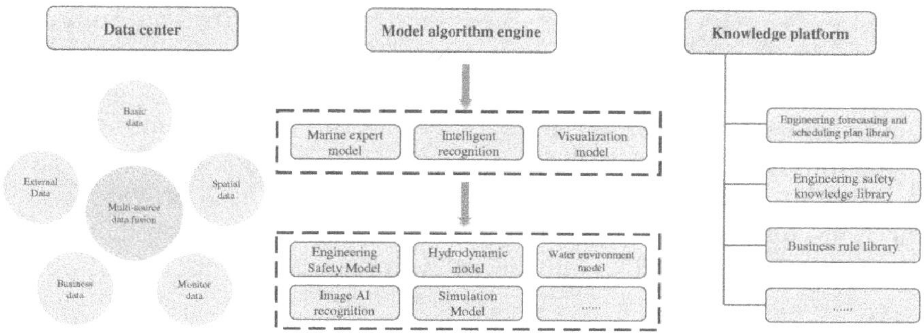

Figure 5.5. Development diagram of digital twin core technology system.

This will provide technical support for establishing smart and high-quality water networks in the Yangtze River Delta. Specifically, the project entails:

Construct a fluorescent array based on nanoclusters to effectively differentiate trace metals and typical dissolved organic matter in water, as well as water samples from various regions.

Develop a portable, real-time spectral monitoring and analysis instrument utilizing a fluorescence sensing array.

Establish a broadband spectral detection system to assess the concentration of suspended sediment in complex water bodies, as well as detect microorganisms, algae, and other aquatic organisms using broadband transmission and scattering spectra.

5.2 Development of Decision Analysis System for the Smart and High-Quality Water Network in the Yangtze River Delta

5.2.1 *Prediction of spatio-temporal distribution of water resources based on spatio-temporal sequence large model*

Drawing upon the framework of intelligent and high-quality water network indicators, we comprehensively utilize domestic and international sources of Earth big data (such as satellite observations, economic and social data, and statistical information) to conduct case studies of smart and high-quality water networks in the Yangtze

River Delta at various scales, including regional, provincial, municipal, and county levels. We aim to develop a basic dataset and algorithm toolkit with independent intellectual property rights that is publicly shareable and capable of meeting the requirements for indicator monitoring and evaluation. This will enhance the construction standards of key hydrological station networks and provide data, methods, and tools to support online monitoring and prediction of the spatiotemporal distribution of water resources in the Yangtze River Delta.

5.2.2 *Research and development of data products and models for online assessment, early warning, and configuration of high-quality water network security risks in the Yangtze River Delta*

Online monitoring and assessment of water network security risks are crucial for ensuring the safe operation of water conservancy projects. It enhances the integration, refinement, and intelligence of management decisions, facilitates communication between key departments, connects various functional units, and enables management personnel to gain a comprehensive understanding of the current situation, thereby improving coordination capabilities. By integrating modern information technology and water network transmission communication technology, we can conduct deep mining and analysis of massive data, extract meaningful key information, and take corresponding actions to address issues, thereby enhancing the scientific rigor and comprehensiveness of decision-making. This will improve water situation monitoring and intelligent scheduling, optimize water resource utilization, reduce costs, achieve sustainable development, and enhance water resource allocation. The following focuses on the research and development of data products and models for online monitoring, assessment, warning, and configuration of water network security risks, including AIoT systems, scheduling algorithms, and the associated challenges.

5.2.2.1 *Current status of the AIoT system for smart water network*

The AIoT system is a comprehensive software platform that integrates monitoring, data processing, intelligent scheduling, and other functions, forming a complete sensing–transmission–knowledge–application closed-loop system. This system enables end-to-end process management from monitoring to scheduling and facilitates dynamic, intelligent allocation of water resources based on various task events. Currently, China is on par with international standards in traditional water resource ecological monitoring, with an increasing variety of equipment and methods. The system typically consists of the following components:

(1) IoT Perception Foundation Layer

IoT sensing facilities are a vital component of smart water management, providing robust information resources to ensure the safety of water supply enterprises' production decisions. Constructing IoT sensing infrastructure involves monitoring pipeline water volume, pipeline liquid levels, pipeline water quality, sewage treatment plants, pump stations, manhole covers, and gates, utilizing real-time monitoring data for construction. To meet the needs of drainage network management, we must improve monitoring capabilities in terms of flow rate, liquid level, and water quality. Water quality monitoring indicators include pH, suspended solids (SS), chemical oxygen demand (COD), ammonia nitrogen, oxidation-reduction potential (ORP) of the solution, and conductivity, specifically as follows:

Drainage pipeline specifications: According to the requirements of GB 50014-2021, "Code for Design of Outdoor Drainage," the maximum design flow velocity for metal pipelines should not exceed 10 m/s, while for non-metallic pipelines, it should not exceed 5 m/s. Furthermore, as per TCECS 869-2021, "Technical Regulations for Online Monitoring of Drainage Pipeline Networks," the measurement error for liquid level should not surpass 1% of the full range, with a measurement resolution of 1 mm. Additionally, the flow measurement error must not exceed 5% of the full range.

pH: One of the crucial hydrochemical indicators, pH is integral to numerous water treatment processes. A low pH level in water can significantly corrode pipeline networks.

Suspended solids (SSs): These include inorganic and organic particles, sediment, clay, and microorganisms that float in water. The concentration of suspended solids serves as a vital indicator of water quality.

Chemical oxygen demand (COD): This comprehensive indicator reflects water quality conditions. Oxidants oxidize organic substances in water, reducing the oxidant's oxygen content. A higher COD value indicates a higher mass concentration of reducing substances (such as organic matter) in the water, which can decrease dissolved oxygen concentration, leading to insufficient biological oxygen, biological death, and degraded water quality with foul odors.

Ammonia nitrogen: An important factor affecting sensory water quality, ammonia nitrogen is a major source of oxygen consumption in water bodies. It primarily originates from human and animal excrement. Excessive ammonia nitrogen can cause water eutrophication and indicate domestic sewage pollution.

Oxidation–reduction potential (ORP): It reflects the macroscopic oxidation and reduction abilities of substances in aqueous solutions. Redox potential is a significant indicator of water body ecological environments.

Conductivity: This physical quantity indicates the ease with which electric charge flows through a substance. Conductivity reflects the mass concentration of metal ions and their corresponding anions in water, indirectly showing the amount of dissolved salt. Increased dissolved salt leads to higher conductivity, which can be toxic to microbial growth.

Sewage plant inflow monitoring: The inlet water quality of sewage treatment plants directly impacts their operational status and is a crucial factor in their operation and regulation. To ensure normal and stable operation and maintain inlet water quality within allowable limits, it is essential to monitor inlet water quality, analyze its characteristics and changes, optimize water treatment plant operational parameters, improve treatment efficiency, and reduce energy consumption. Additionally, real-time monitoring can provide water quality warnings, allowing for timely adjustments to treatment process parameters to ensure effluent quality.

Intelligent pumping station: Install ultrasonic level gauges at the pumping station for real-time collection pit water level monitoring, supporting variable-frequency regulation of the water pump unit. Simultaneously, a current meter at the pipeline network outlet monitors the real-time outflow rate, providing the sewage treatment plant with advance inflow rate information to adjust operating conditions and facilitate scheduling decisions.

Intelligent manhole cover: Leverage wireless networks for digital, networked, and intelligent manhole cover management. The intelligent manhole cover system employs IoT technology to build a management system, enhancing smart water construction data-driven management, improving management efficiency, ensuring safety, and maintaining the normal operation of related equipment.

Control valve: The control gate is equipped with a control cabinet and a pole-mounted monitoring station, uploading data to the computer room's IoT system via a transmission unit. It offers manual and remote control modes, switchable via the control cabinet's conversion switch. Remote detection of equipment status and real-time displays are also available.

(2) IT Infrastructure

IT infrastructure construction includes network transmission, hyper-converged server clusters, data center development, security situational awareness, internet behavior management, next-generation firewalls, operations and maintenance audits (bastion machines), antivirus systems, log audit systems, terminal access systems, and level protection evaluations.

IoT-perceived data are transmitted to the data center through the network, where an eight-server hyper-converged cluster is established. Each server connects to a dual-machine hot-standby aggregation switch, leading to the dual-machine hot-standby core switch. The core switch connects to the dual-machine hot-standby external firewall, with information security devices such as security situational awareness, internet behavior management, operation and maintenance audits (bastion machines), antivirus systems, log audit systems, and terminal access systems attached to provide information security protection services.

(3) Smart drainage command center

The intelligent drainage command center encompasses the construction of the command center facility, interior decoration, and the installation of operation consoles, cabinets, workstations, laptops, and printers.

Electrical system: Based on the requirements for functionality, safety, and intelligent management of the command center, the electrical and lighting systems have been meticulously designed. Adequate redundant capacity is reserved to cater to both short-term and long-term electricity demands. An appropriate number of main maintenance and backup sockets, featuring 220V/16A socket types, have been installed. The cables utilized are low-smoke, halogen-free, flame-retardant copper core and are threaded through steel pipes for protection. The power lines buried beneath the floor are kept separate from computer signal lines, and all distribution lines are routed through metal conduit (groove) for added safety. Furthermore, all power cables meet the relevant fire department requirements as low-smoke, halogen-free, flame-retardant cables. Each circuit in the distribution cabinet is equipped with dual protection against short circuits and overcurrents.

Network cabling system: Fiber optic splicing and cabling are carried out by burying seven-hole pipes in ground trenches for concealed wiring. A dedicated distribution cabinet is positioned between network equipment, serving as a centralized splicing cabinet for optical cables.

(4) Information transmission layer

The information transmission layer serves as a bridge between the system perception layer and the decision analysis layer. Once the system's perception layer acquires ecological information from monitoring equipment, the system's transmission layer utilizes IoT communication technologies, such as the physical layer Modbus, ZigBee, NB-IoT protocol, and the application layer MQTT protocol, to transmit ecological monitoring data and monitoring equipment information (including monitoring terminal latitude and longitude coordinates, running time, and operational health status) in real time to the remote data center. This provides scientific and accurate data to support overall system data mining, equipment operational health monitoring, and ecological management analysis and decision-making.

5.2.2.2 *Water resource scheduling and its intelligent algorithms*

(1) The necessity of unified scheduling in river basins

Water resource scheduling is intimately tied to the development of human society and its practical needs for the development and utilization of water resources. As knowledge levels improve, science and technology advance, and scheduling practices deepen, and human understanding of water resource scheduling continues to evolve. The scope of water resource scheduling has expanded from small regions to large watersheds and even transcends watershed boundaries. The paradigm of water resource scheduling has shifted from a traditional "supply based on demand" model to the current "sustainable development" model, which comprehensively considers the balance between supply and demand. The objective of water resource allocation has also evolved, from a simple allocation to a coordinated consideration of economic, environmental, and ecological needs within the basin and region, accompanied by effective macroeconomic regulation of water resources. The current situation of unified watershed scheduling is outlined as follows:

Unified Scheduling of the Yellow River Water Volume: The Yellow River has pioneered the unified management of major rivers in China, marking a significant chapter in the unified management of water resources within the basin. The Yellow River Basin, birthplace of Chinese civilization, faces challenges due to economic and social development, resulting in water resource shortages and intensifying contradictions between water supply and demand. This has led to river interruptions, overexploitation of groundwater, the shrinking and drying up of lakes, and ecological environment degradation. In 1987, the State Council approved the "Yellow River Available Water Supply Allocation Plan" (hereinafter referred to as the "87 Water Allocation Plan"), which established annual water consumption quotas for the Yellow River and each province (autonomous region) diverting water from the river. Despite this, water interruption issues persisted. In 1997, the Yellow River experienced its most severe flow interruption, with the longest section exceeding 700 km and accounting for about 90% of the length of the lower reaches. The Lijin Hydrological Station recorded 226 days of flow interruption throughout the year, marking a historic high and garnering widespread attention.

In 1998, the former National Development and Reform Commission and the Ministry of Water Resources jointly issued the "Yellow River Water Dispatching Management Measures," authorizing the Yellow River Water Resources Commission of the Ministry of Water Resources to conduct unified dispatching of Yellow River's water volume. On March 1, 1999, the Yellow River Water Resources Commission issued its first dispatch instruction, officially initiating the unified dispatch of Yellow River water volumes. Since then, the Yellow River has undergone a remarkable transformation, from frequent flow interruptions to smooth river flow.

Unified Scheduling of Water Volume in the Heihe River: The Heihe River in the northern part of China is the first inland river in China to implement unified water resource management. Originating from the northern foot of the Qilian Mountains, it flows through three provinces (autonomous regions) of Qinghai, Gansu, and Inner Mongolia. Due to insufficient total water resources, the supply–demand mismatch in the middle and lower reaches of the Heihe River Basin is prominent. Since the 1960s, factors such as population growth, industrial and agricultural development, and insufficient water resources have contributed to a sharp decline in water volume in the lower reaches of the Heihe River, resulting in the drying up of the tail river channel. The West Juyan Sea and East Juyan Sea dried up in 1961 and 1992, respectively.

In 1997, the State Council approved the "Water Allocation Plan for the Main Stream of the Heihe River." In 2000, a unified scheduling measure known as "closed closure and centralized discharge" was implemented. In 2002, water was successfully diverted to the East Juyan Lake, which had been dry for 10 years, and in 2003, water was similarly diverted to the West Juyan Lake. Since 2004, the East Juyan Lake area at the end of the Heihe River has achieved historic and sustained water levels, earning it the moniker of a "green oasis." Additionally, the water volume regulation of the Heihe River has transitioned from emergency regulation to regular regulation. Through unified water management, the water volume in the lower reaches of the Heihe River has gradually increased, initially stemming the tide of ecological deterioration in downstream areas, restoring and enhancing the ecological environment in certain regions, mitigating water conflicts between regions, and yielding significant ecological, social, and economic benefits.

Water Resource Scheduling for Diverting Water from the Yangtze River to Taihu Lake: The Taihu Lake Basin is densely populated, economically developed, and highly urbanized. Since the 1990s, blue-green algal blooms have frequently occurred in Taihu Lake, posing a threat to the water supply safety of surrounding important cities. In accordance with the water resources regulation policy outlined at the Third Working Conference on Water Pollution Prevention and Control of Taihu Lake, held by the State Council in September 2001—which emphasized "treating static water by moving, releasing pollution by cleaning, replenishing dry water with abundant water, and improving water quality"—the Taihu Lake basin has implemented unified regulation of water resources, focusing on the diversion of water from the Yangtze River to Taihu Lake. Since the implementation of the Yangtze River Diversion Project in 2002, by 2022, the project had diverted 36.04 billion m^3 of water from the Yangtze River through the Wangyu River, averaging 1.72 billion m^3 annually. The amount of water flowing into Taihu Lake through the Wangyu River was 16.29 billion m^3, averaging 780 million m^3 annually. Providing 31.45 billion m^3 of water downstream through the Taipu River, averaging 1.5 billion m^3 annually, the project has ensured a reliable and effective water resource guarantee for the regional water environment and ecological security. Therefore, based on historical experience, unified scheduling of water resources is a crucial measure for maintaining the health of rivers and lakes.

(2) The effectiveness of water resource allocation in the new era

Since the 18th National Congress of the Communist Party of China, socialism with Chinese characteristics has entered a new era. General Secretary Xi Jinping has proposed the concept of "water conservation priority, spatial balance, systematic governance, and dual pronged efforts" for water management. These new ideas and emerging situations have set higher standards for water resource allocation work. Water diversion projects and water resource scheduling are arranged in a coordinated manner in terms of space and time, scientifically and reasonably allocating water resources and playing a pivotal role in achieving spatial balance. The 20th National Congress of the Communist Party of China has embarked on a new journey to comprehensively promote the great rejuvenation of the Chinese nation through Chinese-style modernization. To implement Xi Jinping's ecological civilization ideology, uphold the spirit of General Secretary Xi Jinping's important discourse on water management, and build a modern society of harmonious coexistence between humans and nature, it is imperative to implement and enforce the unified dispatch of water resources in river basins. The Party Group of the Ministry of Water Resources has clearly outlined the implementation of the National Water Network Major Project and the restoration of the ecological environment of rivers and lakes as one of six paths for promoting high-quality development of water conservancy in the new stage. Water resource allocation is a vital means of revitalizing the ecological environment of rivers and lakes and is an inevitable requirement for accelerating the development of the national water network.

During this period, basin management agencies and local water administrative departments at all levels have innovated their institutional mechanisms, enhanced their management capabilities, and promoted the high-quality development of water conservation in the new stage. Especially after the 2018 institutional reform, the central government approved the "Three Determinations" regulations of the Ministry of Water Resources, explicitly listing "responsible for water resource dispatch of important watersheds, regions, and major water transfer projects" as one of its primary responsibilities. As a result, water resource dispatch work has been further strengthened, the system has become more robust, the organization has gained more power, and the outcomes have become more prominent, continually creating new opportunities in the field.

(A) The System and Mechanism of Water Resource Dispatch and Management are Continuously Improving

Introduction of Dispatch Management System Documents: Adhering to problem-oriented, goal-oriented, and result-oriented approaches and based on in-depth investigation and research, the Ministry of Water Resources has compiled and issued the "Management Measures for Water Resources Dispatch" to pinpoint prominent issues in water resources dispatch work. Drawing on existing experience in water resource management and in accordance with the relevant requirements of implementing the strictest water resource management system, the document clarifies scheduling objectives from a basin perspective and outlines unified scheduling procedures. It delineates the water resources dispatch and management authorities among the State Council's water administrative department, the basin management agency, and local water administrative departments. Additionally, it establishes mechanisms for scheduling negotiations and coordination, scheduling warnings, ecological water replenishment, information sharing, and other critical processes. In accordance with the systematic requirements for water resource scheduling, the document will coordinate water use for daily life, production, ecological purposes, etc. It clarifies the content and requirements of water resource scheduling management, ranging from the preparation of scheduling plans and annual scheduling plans, scheduling implementation, supervision and inspection, to accountability, thereby promoting the standardization and institutionalization of water resource scheduling management.

Strengthening the Construction of Supporting Systems: In 2022, the initial batch of cross-provincial river basins and major water transfer projects for water resource allocation was identified, encompassing 55 cross-provincial rivers and 11 water transfer projects with clear approval and filing requirements. The "Notice on Further Strengthening the Unified Dispatch and Management of River Basin Water Resources" was issued, outlining requirements to bolster the unified dispatch functions of river basin management institutions and enhance the unified dispatch and management of river basin water resources. Technical guidelines for water resource scheduling plans and annual scheduling plans were developed, improving the technical framework for water resource scheduling. Simultaneously, notable progress has been achieved in the development of water resource dispatch systems at both the river basin and provincial levels, exemplified by the promulgation of various regulations such as the "Implementation Rules for Water Resource Dispatch Management of the Jiang Water Resources Committee of the Ministry of Water Resources," the "Sichuan Province Water Resource Dispatch Management Measures," and the "Inner Mongolia West Liaohe

River Basin Water Dispatch Management Measures (Trial)." Provincial water administrative authorities in Beijing, Hebei, Shanghai, Anhui, Shandong, Henan, and other provinces have also issued water resource dispatch management measures or implementation rules.

Improving the Scheduling Management Mechanism: River basin management agencies and provincial water administrative departments were reorganized to clarify water resource dispatch management agencies and responsible individuals, refining the water resource dispatch organizational system. Mechanisms for consultation, coordination, early warning, ecological replenishment, and information sharing were established and improved. For instance, the Yangtze River Water Conservancy Commission and the Taihu Lake Basin Authority formed a consultation mechanism for basin scheduling, while the Yellow River Water Conservancy Commission, the Huaihe River Water Conservancy Commission, and the Songliao Water Conservancy Commission established an ecological flow early warning response mechanism. The Haihe River Water Conservancy Commission established an ecological water replenishment mechanism, and the Pearl River Water Conservancy Commission enhanced its information sharing mechanism for water resource scheduling. Each provincial-level water administrative department clearly defined the water resources dispatch management organization and the responsible dispatch personnel, further reinforcing dispatch responsibilities. Provincial water administrative departments, such as Hebei, Shandong, Guangdong, and Chongqing, established dedicated water diversion management offices to fulfill their relevant functional responsibilities.

(B) Unified Scheduling and Orderly Promotion of Water Resources

Inter-provincial River Dispatching: Fifty-one inter-provincial river basins, including the Yellow River and Xijiang River, implemented unified water resource scheduling, adhering to a systematic approach and coordinating multi-objective scheduling, yielding significant results. The joint operation of the Yangtze River's main and branch reservoirs effectively responded to the most severe and prolonged meteorological and hydrological drought since 1961. The Yellow River has maintained continuous flow for 24 years, Heihe East Juyan Lake has not dried up for 19 consecutive years, and Taihu Lake has achieved the goal of ensuring drinking water safety and avoiding large areas of black and odorous water for 15 consecutive years. The Beijing–Hangzhou Grand Canal has achieved full water flow connectivity twice, significantly improving the ecological environment of rivers and lakes along its route. The Zhanghe River has been fully connected, providing robust guarantees for water stability and supply safety along the route. The most severe drought in the Hanjiang River Basin in 60 years has been effectively addressed, and the water surface of Baiyangdian Lake remains stable at over $250\,\mathrm{km}^3$, aiding in connecting the entire Beijing-Hangzhou Grand Canal.

Intra-provincial River Dispatching: Provincial water administrative departments determined the list of river basins within their respective administrative regions for water resource allocation, promoting the unified allocation of water resources. As of April 2023, 31 provinces (autonomous regions and municipalities directly under the central government) have issued a list involving 270 rivers. Intra-provincial river scheduling has emerged as a potent lever for promoting the unified scheduling of water resources nationwide.

Standardizing Water Transfer Project Scheduling: A ledger for water diversion projects was established, encompassing 236 large- and medium-sized water diversion projects under construction and serving as the foundational database for industry management. Implementing hierarchical management, eleven major water diversion projects were included in the unified management of the Ministry of Water Resources, with seven completed projects adhering to water resource dispatch requirements. The provincial water administrative department issued a list of confirmed provincial water diversion projects, including 80 water diversion projects under management oversight. Management standards were clarified, steadily advancing the standardized management of water diversion projects, gradually standardizing management work for projects such as the Guangdong Dongshen Water Supply Project, Zhejiang's East Water Diversion Project, and Shandong's Jiaodong Water Diversion Project, and further enhancing dispatch capabilities and levels. Simultaneously, research into improving quality and efficiency was deepened, promoting the sustained and effective utilization of comprehensive engineering benefits.

(C) Ecological Dispatch Effectiveness

Carry out the Mother River Recovery Action: The vitality of the Yongding River Basin has reemerged. Since the initial approval of the water dispatch plan in 2021, the entire line has experienced flow four times, resulting in a significant rise in the groundwater level along the route. The ecological corridor has gradually stabilized, and the sense of fulfillment and well-being among the people living along the river has greatly improved. On September 27, 2021, the Yongding River achieved its first full-line flow in 26 years. By 2022, it experienced two full-line flows in the spring and autumn, reconnecting with the Grand Canal of Beijing and Hangzhou for the first time in a century. The total number of days of full-line flow throughout the year reached 123, with 195 days of water (or ice) present in the entire line. On March 22, 2023, World Water Day, the river once again experienced full-line water flow. The ecology of the West Liaohe River is gradually improving, with the length of the main stream flowing through water increasing year by year. In 2022, the Wobao hub of the General Office witnessed the first water passage since 2002. In the spring of 2023, the water head of the main stream reached approximately 57.15 km below the Wobao hub, achieving a total length of 135.15 km for the main stream with flowing water. This is of great significance in successfully addressing the two major sandy areas of Horqin and Hunshandake. The Zhanghe River has been fully connected to the sea through the Zhangwei New River, with the entire line connected for a total of 89 days as of June 30, 2023.

Carry Out Ecological Dispatch in Key River Basins: Since 2020, the Yellow River has implemented ecological regulation throughout its entire basin, expanding from the main stream to tributaries, from downstream to the entire river, and from the river's interior to its exterior. The ecological flow of key rivers has met the standards, effectively enhancing the ecological corridor function of the river. The main stream of the Yangtze River actively promotes ecological regulation experiments within the basin, including all cascade reservoirs such as Wudongde and Baihetan on the Jinsha River. The Heihe River Basin has implemented measures such as "closing the entire line and centralized discharge," "restricting water diversion," and "flood control" to fully coordinate the use of water for daily life, production, and ecological purposes.

Promote the Improvement of Important Lake Wetland Ecology: The ecological safety of key lakes, such as Taihu Lake and Dianchi Lake, has been ensured through the timely launch of ecological water supplementation projects. Ecological water replenishment has continued to improve the water surface area and wetland ecological environment of Taitema Lake at the tail of the Tarim River. The Wuliangsuhai ecosystem has shown continuous improvement, with biodiversity gradually recovering. The ecological environment in the Heihe River Basin has significantly improved, with the water surface area of East Juyan Lake remaining above 35 square kilometers for many years and the area of Ejina Oasis steadily expanding. The ecological water use in wetlands such as Xianghai and Momoge has been effectively restored. The Yellow River has been introduced into Hebei to supplement water levels and support the ecological water level compliance of Baiyangdian Lake, achieving a 100% level. The Niulan River continues to benefit Dianchi Lake through water replenishment, leading to continuous improvements in water quality. In 2022, the Ningxia Hui Autonomous Region exempted water fees for ecological water replenishment of rivers and lakes and, for the first time, replenished water to the Yinchuan Binhe Wetland and Huinong Jianquan Lake Wetland, effectively supporting the ecological recovery of these water bodies.

5.2.2.3 *Experience and application analysis of water resource scheduling*

Following long-term practical exploration, water resource scheduling has accumulated rich and valuable experience, which should be further summarized and formalized into a systematic understanding.

(1) Adhere to the system concept

The watershed is the most fundamental and distinctive feature of rivers and lakes. It is crucial to adhere to the principle of "overall planning of upstream and downstream, trunk and tributaries, left and right banks." In the unified scheduling of water resources, the Ministry of Water Resources fully leverages the role of top-level planning, design, and comprehensive coordination among all parties. Basin management agencies effectively fulfill the work responsibilities of the "four unifications" of basin governance and management (unified planning, unified governance, unified scheduling, and unified management). Local water administrative departments carry out water resources scheduling guarantee and implementation work in an orderly manner according to relevant task divisions, achieving significant results. Practice has proven that water resource allocation work must adhere to a systematic approach, considering upstream and downstream, main and tributary, left and right banks, as well as domestic, production, and ecological water use systems. It is necessary to firmly establish the concept of "one game" and coordinate the interests of all parties involved.

(2) Persist in following the rules

Rivers are living organisms that adhere to natural laws. By relying on the inherent laws of rivers, we can effectively enhance the efficiency of water resource allocation, maintain the health of river ecosystems, and achieve harmonious coexistence between humans and nature. Taking the Yongding River's water volume regulation as an example, by accurately understanding the natural hydrological laws of the river and continuously refining the water resource regulation approach, a multifaceted regulation pattern has been established. This includes spring multi-source water replenishment, summer flood resource utilization, autumn water replenishment combined with ice storage, and winter ice storage and water conservation. This has laid a solid foundation for ensuring water availability throughout the entire river year-round. We need to continue strengthening the unified scheduling of water resources and research on ecological water replenishment principles, while respecting the fundamental rights of natural river survival and fostering river ethics.

(3) Pursue gradual progress

Implementing unified water resource management, particularly in revitalizing the ecological environment of rivers and lakes, is a long-term endeavor. It necessitates coordinated arrangements and gradual advancements. The Yongding River will achieve full-line flow for the first time in 2021, then twice in 2022, and strive to meet the annual goal of maintaining water throughout the entire length by 2023. The West Liaohe River has, meanwhile, achieved ecological water replenishment for the first time since the Molimiao Reservoir dried up in 2021. In 2022, it witnessed the first water flow from the Wobao Hub of the General Office since 2002, and in 2023, the first dry water head reached the planned urban boundary of Tongliao in nearly 25 years. These developments continue to move toward the overarching goal of gradually restoring full line flow. To align with the actual situation of the watershed, we must set scientific and reasonable scheduling goals, work diligently and continuously promote progress, and ensure the smooth completion of various tasks in water resource management.

(4) Embrace digital empowerment

Water resources allocation is a vital component of strengthening watershed governance and management. It must be based on the watershed as a unit and supported by digital twin technology as a powerful tool. By enhancing forecasting, early warning systems, rehearsal plans, and contingency planning, risks can be detected in advance, warnings issued promptly, plans formulated preemptively, measures implemented swiftly, and scheduling plans dynamically adjusted and optimized to continuously improve scheduling precision.

(5) Intelligent scheduling algorithm

The scheduling algorithm is a cornerstone of the smart water network system, directly influencing the effectiveness of decision-making within the network. Therefore, the primary focus should be on developing scheduling algorithms to ensure optimal water resource allocation and achieve water conservation. The Water Resources Dispatch Center issues instructions to the controller based on the results of these algorithms, which then activate various water equipment to facilitate the dispatch of recycled and clean water resources. By scheduling equipment, the stability of water network operations is ensured, the allocation of different water resources is optimized, water utilization efficiency is improved, and the consumption of clean water resources is reduced, thereby lowering water costs. This realizes intelligent management of dynamic water resource scheduling.

Real-time prediction is the foundation of scheduling, and accurate prediction plays a pivotal role. However, due to the complex variability characteristics of runoff in small- and medium-sized rivers, such as highly nonlinear and multi-time-scale time-frequency sequences, prediction accuracy can be severely impacted. A popular method for addressing this is deep learning, which possesses strong self-learning and adaptive abilities and is well-suited for making predictions. It does not require a prior description of the mathematical relationship between input and output but only stores the mapping pattern. By setting weight thresholds for forward propagation, if the training data output deviates significantly from the actual value, the weight threshold parameters are continuously adjusted through backpropagation. According to the gradient descent principle of the error function, the sum of squared errors across the entire neural network is minimized, achieving the expected output value through these two processes.

The triggering events for water network scheduling stem from various data sources, including economic development levels, policy and regulatory changes, and government demands. The development of this integrated scheduling system will become a crucial safeguard for water network security in the Yangtze River Delta region. It can not only promptly detect and assess potential risks but also quickly and intelligently allocate resources to ensure the rational distribution and utilization of water resources.

5.2.2.4 *Opportunities and challenges*

(1) Accelerating the national water network project to provide new opportunities for the construction of a national smart water network

This section highlights the potential opportunities presented by the National Water Network Project for advancing the construction of a national smart water network. In order to comprehensively enhance water security, General Secretary Xi Jinping has elevated the construction of the national water network to a national strategic level, requiring the coordination of stock and increment, strengthening interconnection, and accelerating the construction of the main framework and arteries of the national water network. In order to implement the Fifth Plenary Session of the 19th Central Committee of the Communist Party of China on the development of major national water network projects, and conscientiously uphold the important spirit of General Secretary Xi Jinping at the symposium on the high-quality development of the follow-up projects of the South to North Water Diversion Project, the Party group of the Ministry of Water Resources has designated the implementation of the national water network project as one of six implementation paths to promote high-quality water conservation in the new era. The 14th Five Year Plan for Water Security requires promoting the transformation of traditional water conservation projects into new water conservancy infrastructure, accelerating the intelligent transformation of existing projects, and continuously improving the intelligent management of their construction and operation. The Ministry of Water Resources has issued documents such as the "Guiding Opinions on the Implementation of Major National Water Network Projects" and the "Implementation Plan for Major National Water Network Projects during the 14th Five Year Plan Period," which require accelerating the construction of a "complete, safe, reliable, intensive, efficient, green and intelligent, smooth circulation, and orderly control" national water network, strengthening integration with new infrastructure construction, focusing on digital twin projects, promoting intelligent construction, improving digitalization, networking, and intelligence levels, enhancing the ability to forecast, warn, rehearse, and plan, and improving the level of water resource allocation project scheduling and management.

(2) The implementation of the strategy of building a strong cyber nation puts forward new requirements for the construction of a national smart water network

General Secretary Xi Jinping places great importance on building a strong cyber nation and digital China, and has put forward a series of clear requirements for digitization, networking, and intelligence. The 14th Five Year Plan for National Economic and Social Development and the 2035 Long-Range Objectives clearly state the construction of a smart water conservation system to enhance water situation monitoring and intelligent dispatching capabilities on a basin-by-basin basis. The National 14th Five-Year Plan for the Construction of New Infrastructure clearly proposes to promote the construction of digital twins, intelligent simulations, and intelligent business applications for major rivers. The "14th Five Year Plan for National Informatization" mentions multiple aspects of smart water conservancy construction, such as the "Green, Smart and Ecological Civilization Construction Action" chapter, proposing to "promote smart water conservancy construction," requiring "promoting the integration of new generation information technology and water conservancy business, improving the monitoring system of major rivers, strengthening the application of water conservancy big data, accelerating the orderly sharing of important water conservancy data such as hydrology and water resources, and improving the ability of water situation monitoring and intelligent scheduling on a basin by basin basis"; the chapter on "Building a New Perception Infrastructure for IoT Data Communication" regards the digitization and intelligent upgrading of water conservancy public infrastructure as a key component of that infrastructure for IoT data communication; The chapter on "Improving the level of data resource development and utilization" considers the water conservancy field as one of the supporting scenarios for standardized data development and utilization, aiming to enhance the value of data resources; the chapter on "Building an Integrated Intelligent Public Safety System" considers water conservancy facilities as one of the areas to enhance operational situational awareness and intelligent analysis to improve the ability to identify and forecast public safety risks; and the chapter on "Improving and Upgrading Rural Infrastructure" regards the digital and intelligent transformation of water conservancy infrastructure as a critical undertaking.

(3) Digital transformation and localization substitution pose new challenges to the construction of national smart water networks

Currently, China has built the world's largest, most extensive, and most beneficial water infrastructure system. In recent years, with the construction of large-scale water conservancy and hydropower projects such as the Three Gorges Dam on the Yangtze River, the Middle Route of the South to North Water Diversion Project, and the East Route of the South to North Water Diversion Project, China has gained world-leading engineering planning and design technology, mechanized construction methods, and modern construction and operation management capabilities. However, overall, due to technological and economic constraints, the intelligence level of most of China's water conservancy infrastructure is relatively low, and the reliance on foreign brands of information facilities and equipment presents safety hazards.

(4) The construction of digital intelligence lags behind and is far from meeting the requirements of digital transformation

Currently, there is a serious problem of decentralization and fragmentation in the information system of water network engineering, and an integrated operation and management model has not yet been formed. Serious data silos, ineffective utilization of data resources, and an inability to use data for management and decision-making are also present. The level of platformization and intensification is low, with prominent problems of small and scattered, constructions, insufficient platformization and specialization capabilities, and overall low investment efficiency. The level of intelligence is insufficient, with shortcomings in areas such as comprehensive perception, refined control, and intelligent decision-making. Prediction, rehearsal, and predictions based on digital intelligence have not yet been realized; scheduling efficiency is low, and the capabilities of unified, fine, and multi-level linkage scheduling are insufficient, which is not compatible with the construction of the Yangtze River Delta water network in the new era, achieving cross-domain collaborative scheduling and efficient operation.

(5) Key equipment heavily relies on foreign countries, and independent and controllable domestic substitution is a long way to go

The important network facilities and information systems within the water conservancy industry are listed as national key information infrastructure, requiring monitoring, defense, and mitigation of network security risks and threats originating from both domestic and foreign sources, and safeguarding this infrastructure from attacks, intrusions, interference, and destruction. At present, China's core components heavily rely on foreign countries, and we need to firmly grab the bull by the horns through independent innovation in core technology, seize breakthroughs in cutting-edge network technologies and key core technologies with international competitiveness, accelerate the promotion of domestic independent and controllable substitution plans, and build a secure and controllable information technology system. These core business systems play an important role in supporting and ensuring the security and stability of the national economy, people's livelihoods, and economic operations. Therefore, higher requirements need to be put forward for alternative high stability, high availability, and high real-time performance. According to statistics, in the informationization and industrial control equipment of water conservation facilities, such as the East Route of the South to North Water Diversion Project, the average proportion of foreign equipment applications exceeds 90%. Among them, the proportion of American equipment exceeds 80%, and the proportion of domestically produced equipment applications is less than 10%, posing risks such as technological bottlenecks, supply chain disruptions, and uncontrolled information security.

(6) Water network information perception model based on the integration of air, space, earth, and sea

The latest smart water network requires the integration of multiple monitoring and information systems, such as satellite remote sensing technology, unmanned aerial vehicles, manual river patrols, and information platforms, to regulate the water network. "Air" refers to low-altitude unmanned aerial vehicle monitoring, which utilizes unmanned aerial vehicles and photogrammetry technology to obtain various geographic information of key areas. "Sky" refers to remote sensing interpretation of water networks through high-resolution image data, while "Earth" refers to ground monitoring and on-site verification. The primary challenge is the ease of misjudgment in manual on-site identification; for this, an information system establishes a remote sensing supervision platform integrating aerial, space, and terrestrial data. The "integration of air, space, and earth" regulatory model fully combines the advantages of the three regulatory methods of "air," "sky," and "earth," and addresses their respective shortcomings, providing effective data for the smart water network. The specific technical solution for the "integration of air, space, and earth" regulatory model is mainly divided into the following three parts:

(A) Remote sensing interpretation collects high-resolution images of the target area through satellite remote sensing platforms such as domestically produced Gaofen-2 and Beijing-2. Preprocess the image data, combine legal information and boundary ranges to form a monitoring work base map of the target area and establish interpretive markers.
(B) A preliminary list of suspected issues is formed by combining unmanned aerial vehicle (UAV) aerial photography and manual on-site reviews in key areas using remote sensing interpretation. UAV aerial photography and a manual on-site review are conducted on key suspected areas. Based on the review results and feedback from the river management department, necessary corrections and improvements are made to the spatial characteristics and attribute information of the problems discovered in remote sensing interpretation and uploaded to the remote sensing supervision platform for river and lake shorelines.
(C) The remote sensing supervision platform adopts remote sensing interpretation technology, 3S technology, etc. We have developed

a remote sensing monitoring platform for river and lake shorelines, established a remote sensing monitoring database for river chiefs, and achieved effective management of basic information, remote sensing information, on-site reviews, and related information. This platform can be used to provide feedback on problem reports to the local river and lake management department. The local river and lake management department conduct on-site reviews and rectification; they then upload the results to the remote sensing supervision platform for review by higher-level management, forming a comprehensive process supervision model encompassing satellite remote sensing interpretation, unmanned aerial vehicle aerial photography of key areas, manual on-site review, online problem distribution, offline rectification, and online review.

5.2.3 *Online monitoring, evaluation, and strategic response of the impact of extreme weather events and emergencies on smart and high-quality water networks*

An online monitoring, evaluation, and strategic response mechanism needs to be established to address the impact of extreme weather events and emergencies on smart and high-quality water networks. This mechanism will enhance the adaptability of water resource systems to the increasing normalization of extreme weather events. Moving forward, it is imperative to further refine the reliability and adaptability of long-term extreme event forecasting: Clarify the incubation, occurrence, development, and mutual feedback mechanisms of typical extreme events affecting water resource security in the Yangtze River Delta, viewed through a lens of multi-level interactions. Decouple the effects of external forcing and internal variability in the climate system on extreme events, emphasizing the elucidation of contributions and mechanisms of external forcing, such as human activities, and revealing the potential impacts of clean air policies. Clarify the sources of uncertainty in extreme event estimation across different future time scales. Develop corresponding new technologies and data products for extreme event and sea level observation constraint estimation, thereby improving the reliability of these estimates.

5.3 Application Services of Smart and High-Quality Water Networks in the Yangtze River Delta

5.3.1 *Dynamic detection and prediction of water quality and quantity control and dispatch demand in the Yangtze River Delta integration*

The current status and future prospects of social and economic development in the Yangtze River Delta region must be elucidated, along with the application needs for local and inter-provincial water quality control and scheduling within the context of smart and high-quality water networks. This involves analyzing the current socio-economic landscape of the region and predicting its development over the next 5, 10, 20, 30, and 50 years. The application needs of smart and high-quality water networks in future scenarios are assessed.

First, an analysis of population and administrative divisions, socio-economic characteristics, and the construction, supply, and overall design of smart and high-quality water networks in the Yangtze River Delta is conducted. Based on the objectives of meeting water quality and quantity requirements, ensuring ecological sustainability, and aligning with government initiatives, this study examines the population, administrative divisions, socio-economic characteristics, and the demand for smart and high-quality water network construction in the region. It also explores the micro-storage and micro-circulation of regional water resources. Consequently, a comprehensive design scheme for the Yangtze River Delta smart and high-quality water network risk perception and analysis system is formulated.

5.3.2 *Construction of a channel condition perception network based on the smart and high-quality water network in the Yangtze River Delta*

This involves exploring smart waterways by leveraging the construction of high-quality water network resources in the Yangtze River Delta. Through an in-depth analysis of the distribution, utilization, and planning of the region's water network, combined with its unique characteristics, this study highlights the positive role of the waterway condition perception service network in promoting efficiency gains in waterway planning, construction, management, maintenance, and other aspects. Initially, we establish a perception system through strategic planning to break down existing data barriers and minimize redundant data collection efforts. Secondly, a perception and communication network is built for the navigation function of the water network to collect and transmit basic waterway data. Thirdly, digital waterways are developed by integrating the navigation functional elements of the water network into the urban metaverse. Waterway condition data are mined to support waterway planning, construction, management, and maintenance decisions. This will foster the intensive, effective, and comprehensive development and utilization of water network resources in the Yangtze River Delta.

5.3.3 *Construction and operation of the smart and high-quality water network security risk perception and comprehensive monitoring, early warning, and forecasting platform in the Yangtze River Delta under the normalization of extreme weather conditions*

Remote sensing and GIS technology can extract the spatial distribution characteristics of urban and rural residents near major water sources in the Yangtze River Delta. A comprehensive analysis of the spatiotemporal changes should be conducted, coupling relationships and matching differences in flood control, irrigation, urban water use, industrial water use, power generation, ecological water use, and inter-provincial and inter-city water transfer demand within these water sources. The evolving trends in water demand matching within the Yangtze River Delta are then predicted, considering the influence of multiple factors, including climate change and human activities. This is followed by integration and assimilation of multi-source remote sensing big data, online monitoring data, model data, and artificial intelligence to construct high-quality water network safety risk intelligent perception models and early warning and forecasting models. Corresponding comprehensive platforms are developed to support these efforts.

Chapter 6

Estimating Terrain Elevations at 10 m Resolution by Integrating Random Forest Machine Learning Model and Remotely Sensed Data from ICESat-2, Sentinel-1, and Sentinel-2 Satellites

Accurate mapping of terrain elevations at a large scale and fine resolution can characterize detailed surface height and geomorphic changes and is highly critical for the studies of the internal motions and external forces of the Earth. The emergence of the Ice, Cloud, and Land Elevation Satellite-2 (ICESat-2) offers unprecedented opportunities for global elevation mapping with high vertical accuracy using three-dimensional photon points. However, the ICESat-2 photon points remain sparse in terms of spatial/horizontal resolution, making it unable to satisfy the high-resolution demands of terrain elevation mapping and digital elevation model production. A few previous studies have attempted to estimate elevations/topography in regions with single landscapes and land cover (e.g., forests, shallow water, and polar regions) by combining ICESat-2 data with other passive satellite remotely sensed data. However, the potential and capability of ICESat-2 for mapping elevations across spatially continuous, large regions with multiple complex land cover types remains unknown. In this study, a spatially continuous, large-scale terrain elevation estimation method is developed under multiple land covers based on the random forest model and freely available satellite data from ICESat-2, Sentinel-1, and Sentinel-2. The core principle is to construct a random forest model that can characterize the complex relationships between the ICESat-2 ATL03 terrain elevations and their corresponding land-cover-related polarization characteristics and spectral variables derived from Sentinel-1 and Sentinel-2, respectively. Integrating the superiorities of the data from these three different sensors enables the proposed method to extrapolate terrain elevations with decimeter-level vertical accuracy and 10 m spatial/horizontal resolution simultaneously, without prior *in situ* data or manually set parameters. The proposed method is tested using elevations from 2021 to 2022 at the third largest island (Chongming Island, Shanghai) in China. The estimated terrain elevations are locally validated with airborne LiDAR-derived elevations.

Moreover, they are compared with the ICESat-2 ATL08 height_terrain_bestfit data and the Global Ecosystem Dynamics Investigation L2A elev_lowestmode data from the global perspectives. The predicted elevations exhibit a high correlation with the measured elevations from the two airborne LiDAR validation regions, with root mean square errors (RMSEs) of 0.34 and 0.59 m. The average RMSEs of the predicted elevations at different land covers are 1.26 and 1.18 m when compared with those derived from ATL08 and GEDI L2A, respectively. No remarkable abnormal predicted elevations are observed. This finding suggests the satisfactory robustness performance of the proposed method under different land covers and relatively good consistency between the predicted elevations and the actual terrain of the entire island. As far as we know, this work is the first to map elevations at 10 m resolution using only newly available satellite active and passive remotely sensed data, without ground truth surveys, manual intervention, or prior knowledge. Differing from existing studies that considered terrain elevation mapping only at single land covers, the proposed method demonstrates the capability and effectiveness of ICESat-2 for any landforms and land covers and shows great potential for high-accuracy and high-resolution time-series terrain elevation estimation and updating at regional, national, and global scales.

6.1 Introduction

Terrain elevations can characterize the detailed surface undulations and variations, as well as the internal structures and motions of the Earth (Bespalko and Hutsul, 2021; Cooper *et al.*, 2013; Iwahashi *et al.*, 2018). An accurate mapping of terrain elevations allows for a scientific topographic and geomorphological understanding of the Earth's surface and provides a scientific basis for resource management, environmental monitoring, urban planning, natural disaster assessment, and policy formulation (Ahmad *et al.*, 2018; Habtu and Katihally, 2018). Despite existing global digital elevation products being well-established and widely used, they suffer from poor data resolutions and unacceptable errors or untimely and constant updating frequencies (Liu *et al.*, 2007; Liu, 2008). Traditional field surveying methods, such as leveling and real-time kinematic (RTK), can obtain highly accurate elevation data with centimeter or even millimeter accuracy. Nevertheless, these methods are highly time-consuming and labor-intensive and can be applied discretely and locally only in accessible regions (Starek *et al.*, 2021; Wulder *et al.*, 2012).

Remote sensing technologies on various platforms enable local to global Earth observations from a noncontacting perspective. Passive and active remote sensing data are effective data sources for directly measuring or indirectly deriving terrain elevation with varied spatial coverage, data accuracy, resolution, and temporal frequency (Nelson *et al.*, 2009). Passive remote sensing technologies do not directly record elevation. As such, some strategies or methods such as oblique photography, stereo image pairs, and waterline detection are needed for elevation derivation with the assistance of ground truth observation data. In particular, stereo satellites such as SPOT, ALOS, and WorldView, which utilize multispectral bands, offer unique advantages in estimating terrain elevations globally using self-generated along-track or across-track stereo image pairs. However, passive remote sensing technologies are vulnerable to environmental conditions (e.g., shadows, lights, occlusions, and clouds). Moreover, they do not provide good penetration performance in forested and water environments (DeWitt *et al.*, 2015). Interferometric synthetic aperture radar, a representative active remote sensing technology, provides the possibility to retrieve large-scale terrain elevations even in harsh environmental conditions; however, the signals suffer from saturation effects and are greatly affected by opaque objects and terrain undulations, particularly in complex landscapes (Mohammadi *et al.*, 2020). Additionally, elevations of the turbid waters, dense forests, and mountainous regions cannot be satisfactorily obtained by this technology.

Light detection and ranging (LiDAR) can actively emit and receive laser pulses and can directly obtain high-precision three-dimensional (3D) measurements, including elevation. Given the advantages of superior penetration capability, low environmental dependence, and high-precision observation, LiDAR has boomed in mapping terrain elevations at different scales by various platforms (Swatantran *et al.*, 2016). Terrestrial and airborne platforms can generate dense 3D point clouds to acquire detailed terrain elevations; however, they are greatly limited by large-scale spatial coverage, low economic cost, and the need for time-series observations (Markus *et al.*, 2017). By contrast, spaceborne/satellite LiDAR exhibits outstanding potential in regional and global-scale elevation mapping due to its large spatial coverage and high temporal frequency. Ice, Cloud, and Land Elevation Satellite-2 (ICESat-2) is a newly advanced spaceborne LiDAR (Neumann *et al.*, 2019). The core instrument is the Advanced Topographic Laser Altimeter System (ATLAS). This new photon-counting laser sensor shows great potential for mapping global vegetation characteristics, shallow water bathymetry, and land topography with submeter accuracy (Lian *et al.*, 2022).

ICESat-2/ATLAS offers a total of 22 standard data products, of which the ATL03 contains the ellipsoidal height/elevation, longitude, and latitude of each photon event (Ranndal *et al.*, 2021). With high-accuracy performance, the elevation information from ATL03 data has been used as ground truth data or control points for validating or generating products from other remotely sensed satellite data in various applications (Brunt *et al.*, 2019; Li *et al.*, 2022; Yap *et al.*, 2019). The along-track sampling spacing is 0.7 m. However, the footprints are relatively sparse in the across-track directions (3 km). Thus, the spatial/horizontal resolution cannot accurately reflect detailed terrain elevations. Integrating ICESat-2 with other available high-resolution satellite imagery data is a promising solution to realize the full potential of different techniques in addressing the challenges and inadequacies of temporal, spatial, and spectral resolutions. Sentinel-1 (synthetic aperture radar imagery at C-band) and Sentinel-2 (multispectral imagery) are publicly available satellite data with high temporal (5 d) and spatial (10 m) resolutions. Therefore, the insufficiencies in the spatial resolutions of ICESat-2 can be compensated if effectively integrated with the series datasets of Sentinel. However, these three datasets are heterogeneous in terms of resolution, dimension, and connotation. Integrating multisensor heterogeneous data is a great challenge for traditional fusion methods.

A new method is developed for terrain elevation estimation based on the random forest model and a joint use of the ICESat-2, Sentinel-1, and Sentinel-2 data in the current study. According to the basic principle indicating that elevations differ with land cover and that land cover exhibits varied polarization and spectral characteristics, the random forest technique is used to model the nonlinear and complex relationships between ICESat-2 elevations and the other two Sentinel datasets. Based on this model, polarization and spectral variables related to land cover types from Sentinel-1/Sentinel-2 imagery are then constructed at 10 m resolution to predict/extrapolate the elevations of the regions uncovered by ICESat-2. No *in situ* or ground truth data are required in the proposed method because ICESat-2 ATL03 data act as prior elevation information (a seed elevation). Sentinel-1 and/or Sentinel-2 polarization and multispectral metrics are effective covariables in the global-level mapping of tidal flat topography, forest canopy height, and aboveground biomass (Silva *et al.*, 2021; Xu *et al.*, 2022; Guerra-Hernández *et al.*, 2022; Xi *et al.*, 2022; Li *et al.*, 2020) when coupled with the ICESat-2 data. Nevertheless, whether the integration of the three datasets shows equivalent capability and effectiveness in large-scale terrain elevation mapping remains unknown, particularly in a large region with different land covers. As far as we know, this work represents the first attempt to demonstrate the capability for mapping large-scale terrain elevations by integrating the ICESat-2 with the Sentinel series datasets.

6.2 Study Site and Datasets

6.2.1 *Study site*

The third-largest island (Chongming Island, Shanghai) in China, with an area of approximately 1,300 km^2, is selected as the study site (Figure 6.1). According to the government plan, Chongming Island will be built into a world-class ecological island. Chongming Island

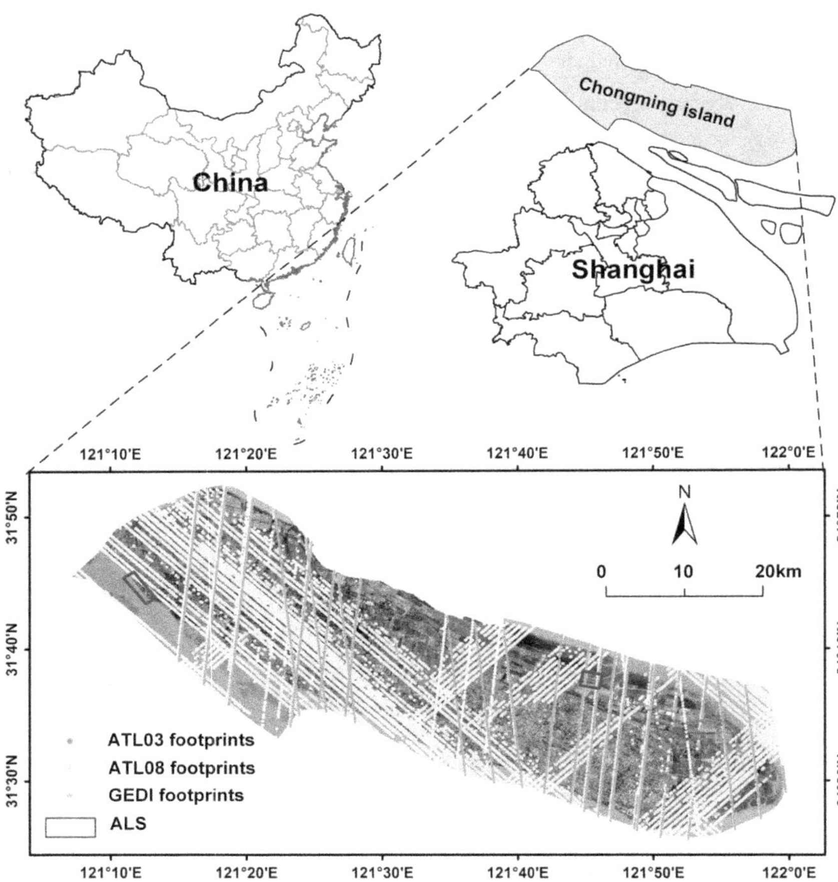

Figure 6.1. Sentinel-2 true-color composite imagery of Chongming Island from Google Earth Engine, overlaid with ICESat-2 ATL03 track (green points, 1659519), ATL08 track (yellow points, 5957), GEDI track (gray points, 27114), and ALS boundaries (red rectangles). The boundaries of Chongming Island are based on the administrative district vector map from Golder Maps, China.

is an ecological barrier for Shanghai, which is of great value for the ecological security of the Yangtze River Delta. The island's topography is flat, with an overall trend showing that the terrains in the northwest and central parts are slightly higher and that those in the southwest and eastern parts are slightly lower. Chongming Island has highly developed inland water systems and a very high vegetation coverage rate, which poses two huge challenges for terrain elevation estimation using remotely sensed data. The land cover is rich, mainly constituted by tree cover, grassland, bare land, cropland, shrub, built-up areas, water bodies, and wetlands. Chongming Island shows representativeness in terms of ecological value, land cover types, and terrain complexity. A small area in the northern part of Chongming Island belongs to Jiangsu Province, China, and this part was removed from this study.

6.2.2 *Data sources*

6.2.2.1 *ICESat-2*

ICESat-2 data are divided into Levels 0, 1, 2, 3A, and 3B, which include a total of 22 products from ATL00 to ATL23, except ATL05 and ATL22. Detailed information on the product number, spatiotemporal range, and resolution can be found on the official website. ATL03 is a Level 2 product, which corresponds to the global geolocated photon and serves as the basic data for generating other series of products (e.g., ATL06, 07, 08, 10, 12, and 13). The height and geographic coordinates are based on the ITRF2014 frame and the WGS84 ellipsoid, respectively. The six strips of the ATL03 data are segmented according to a distance of 20 m along the track, and the photon points within each segment contain longitude, latitude, and ellipsoidal height information (Neumann *et al.*, 2020). A denoising algorithm is utilized in ATL03 to classify raw photon points into four grades, i.e., low-, medium-, and high-confidence signals, and background noise. The ATL03 data from October 2021 to October 2022 across the entire study site were used. To ensure acquisition of the ground points, we filtered ICESat-2 footprints based on the criterion that only the high-confidence signal photon points (signal_conf_ph of 4) were selected to minimize the influence of background noise and nonground targets (e.g., trees and buildings) on terrain elevation prediction.

ATL08 data (Level 3A) are produced based on ATL03, with the removal of background noises (Neuenschwander *et al.*, 2019). The remaining signal photons are classified into vegetation canopy tops, ground, and canopy photons. The elevation parameters of the ground photon points include the best-fit, mean, maximum, and minimum ground elevations. In this study, the best-fit ground elevation parameter, h_te_bestfit, with high accuracy, was selected for a global quantitative comparison of predicted elevations under different land covers (Tian and Shan, 2021; Urbazaev *et al.*, 2022; Zhao, 2022).

6.2.2.2 *Sentinel data*

Sentinel-1 is equipped with a C-band SAR sensor. It can always provide continuous observations in all weather conditions. Sentinel-1 sensors can work in four scanning modes: SM (island), IW (land and coast), EW (marine area), and WV (vast ocean). Two kinds of alternative polarization modes, including single (VV or HH) and dual-polarization modes (VV+VH or HH+HV), are provided for Sentinel-1 sensors. The study utilized IW scanning mode data with VV+VH polarization, where V and H mean vertical and horizontal polarizations, respectively.

Sentinel-2 is equipped with a multispectral imager (MSI). The MSI comprises 13 band channels from visible light to shortwave infrared (B1–B12 and B8A). The spatial resolutions vary at 10 m, 20 m, and 60 m for different bands. Since December 2015, Sentinel-2 data (Level 1C product) has been officially available for free download to users worldwide. We resampled the Sentinel-2 data (Level 2A) at 10-meter resolution to eliminate variations in spatial resolution.

The Sentinel data were downloaded from Google Earth Engine and processed using the sen2cor software and the Sentinel-1 toolbox. The mean image composites (annual mean composites) of both Sentinel-1 and Sentinel-2 from October 2021 to October 2022 were used to ensure temporal consistency with the ICESat-2 data.

6.2.2.3 *ALS data*

Airborne LiDAR (i.e., airborne laser scanning, ALS) data were used as verification data for estimated terrain elevations due to their high-accuracy performance in deriving terrain elevations. The LiDAR data were collected over two validation regions in the Dongtan and Xitan Wetlands of Chongming Island (red rectangles in Figure 6.1) at low tide. The validation region in Xitan Wetland is a natural and artificial composite ecosystem with an area of approximately 2 km^2, including a wide variety of different target species and land covers, e.g., buildings, roads, mudflats, trees, shrubs, and grasses. The point cloud data for the region were collected on August 17, 2020, using a ZR-M66 unmanned aerial vehicle (UAV) equipped with a Hummingbird Genius LiDAR system. The scanning field of view of this instrument is $360° \times 30°$, and the emitted laser is an infrared laser (905 nm). The flight height and speed were 100 m and 7 m/s, respectively. Approximately 65 million points were finally obtained. In contrast, the validation region in Dongtan Wetland, with an area of approximately 3 km^2, is a tidal flat. The target species and land covers are relatively distinct. However, this region is covered by very dense salt marsh vegetation and several tidal creeks. The tidal flat was scanned by the DJI L1 LiDAR equipped on an M300RTK UAV on March 11, 2023. The scanning field of view was $70.4° \times 4.5°$, and the emitted laser was an infrared laser (905 nm). The flight height and speed were 100 m and 7 m/s, respectively. Approximately 970 million points were obtained.

Taking the RTK measurements at the site as a reference, the vertical and horizontal accuracies of the acquired point clouds for the two validation sites were approximately 0.07 and 0.10 m, respectively. The coordinate system of the point cloud data (x and y) was WGS84, and the elevation (z) was the geodetic height. The collected original point cloud data included nonground points; therefore, cloth simulation filtering was adopted to remove the nonground points (Zhang *et al.*, 2016). In some local regions, manual filtering was complementarily conducted to ensure a satisfactory acquisition of the ground points. Finally, only approximately 26 and 118 million ground points were reserved for verification due to the dense vegetation coverage at the two sites (Figure 6.2).

6.2.2.4 *ESA WorldCover 2020 v100 and Global Ecosystem Dynamics Investigation (GEDI) L2A*

WorldCover 2020 v100 (10 m resolution) is a global land cover product (Zanaga *et al.*, 2022). Eleven types of land cover, namely tree

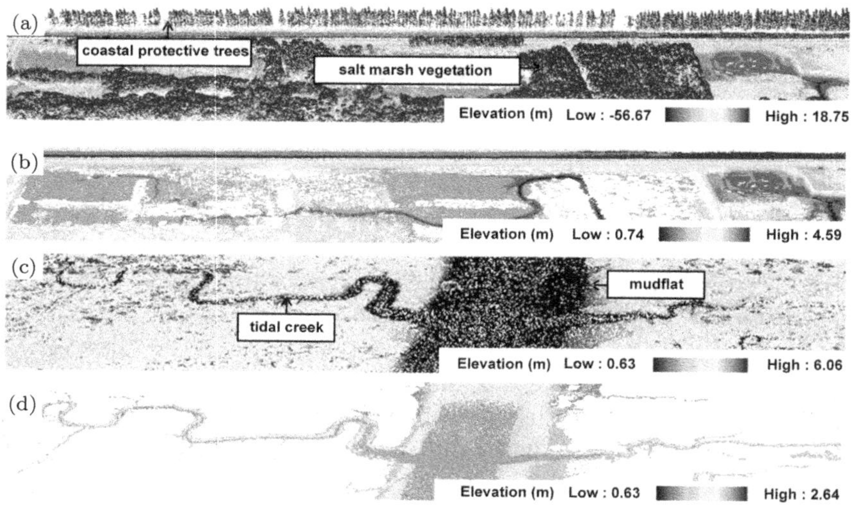

Figure 6.2. Local comparisons of ALS data before and after filtering. (a) and (c) are the raw point cloud data of the validation regions in Xitan and Dongtan Wetlands, respectively. (b) and (d) are the corresponding point cloud data (ground points) after filtering.

cover, snow and ice, grassland, shrubland, cropland, built-up, bare land/sparse vegetation, herbaceous wetland, permanent water bodies, moss and lichen, and mangrove, are contained in the WorldCover 2020 v100 data. In the study site, eight land cover types were observed, excluding snow and ice, mangrove, and moss and lichen. In addition, no obvious height difference was observed between the grassland and shrubland. Consequently, the grassland and shrubland were merged as low vegetation in this study.

GEDI is a joint mission between NASA and the University of Maryland. GEDI makes precise measurements of canopy vertical structure, forest canopy height, and surface elevation using the instrument's three lasers, which produce eight parallel tracks of observations. The along-track and across-track resolutions are 60 and 600 m, respectively. GEDI includes L1, L2, L3, and L4 level products, where the L2 level product is extracted from the L1 waveform data. The GEDI L2 level product includes L2A and L2B, which were publicly released in 2020. Together with ICESat-2 ATL08

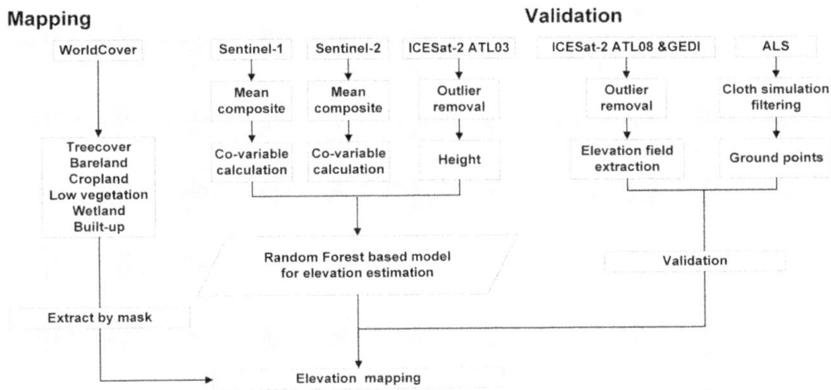

Figure 6.3. Workflow of terrain elevation mapping by the proposed method.

height_terrain_bestfit data, the GEDI L2A elev_lowestmode data from October 2021 to October 2022 covering the study site were selected for a global quantitative comparison of the predicted elevations under different land cover types.

6.3 Methodology

In the proposed model, independent variables include polarization characteristics and spectral variables derived from Sentinel datasets. By contrast, ICESat-2 photon heights are used as dependent variables. The overall workflow is shown in Figure 6.3. The WGS84 ellipsoidal height was used for all the original datasets, which was subsequently converted to orthometric height, referring to the EGM2008 via the GAGE website (unavco.org), and the WGS84 coordinate system and UTM projection (zone 51N) were used.

6.3.1 *Feature selection*

A total of 35 different features were constructed from Sentinel-1 and Sentinel-2 data (Table 6.1). The SAR backscatter coefficients are highly related to surface physical properties and terrain structures.

Table 6.1. Quantities and variables derived from Sentinel datasets.

Data	Variables	Details	References
Sentinel-1	GLCM with a 3×3 kernel for VV and VH bands	Mean Variance Homogeneity Contrast Dissimilarity Entropy Second moment Correlation	(Haralick *et al.*, 1973)
Sentinel-2	Sentinel-2 bands	B1 $\sim$ B8, B8A, B9, B11, B12	
	Sentinel-2-derived spectral indices	NDVI GNDVI SAVI NDWI MNDWI NDBI	(Rouse *et al.*, 1973) (Gitelson *et al.*, 2003) (Huete, 1988) (Gao, 1996) (Allen and Pavelsky, 2018) (Zha *et al.*, 2003)

Therefore, the corresponding texture features from Sentinel-1 VV and VH backscatter coefficients were selected. The gray-level co-occurrence matrix (GLCM) was adopted to derive the texture features using an optimal 3×3 kernel. Eight statistical measures, namely mean, dissimilarity, homogeneity, variance, entropy, correlation, contrast, and second moment, were obtained from the GLCM for VV and VH polarization information. The terrain elevation strongly correlates with land cover. The Sentinel-2 data with multispectral information have a strong ability to recognize land cover details. In addition to the 13 raw band characteristics of Sentinel-2, six types of spectral indices were constructed: Normalized difference vegetation index (NDVI) contributes to vegetation identification; green NDVI (GNDVI) shows the greenness of the land surface; soil-adjusted vegetation index (SAVI) measures land cover; normalized difference water index (NDWI) and modified NDWI (MNDWI) are related to water bodies and wetlands; and normalized difference built-up index (NDBI) distinguishes between urban built-up areas and natural surfaces.

6.3.2 *Random forest*

Random forest shows advantages in dealing with high-dimensional data, removing outliers, and avoiding overfitting (Belgiu and Drăguţ, 2016; Breiman, 2001). Multiple decision trees are adopted in random forest to train and predict samples. The random forest package in R Studio was used in this study (R Core Team, 2015). The 35 features in Table 6.1 were considered independent variables, and the photon heights in ICESat-2 ATL03 were used as dependent variables, with all the variables sampled to a 10 m resolution. A total of 25,687 high-confidence signal photon points (with a signal_conf_ph of 4) in ICESat-2 ATL03 of the study site were selected; these points were arbitrarily segmented into accuracy validation and model training points in a ratio of 3:7. Two critical parameters, *mtry* and *ntree*, were contained in the random forest model, where *mtry* is the number of features used in each individual decision tree and *ntree* stands for the number of regression trees. In this study, *ntree* and *mtry* were assigned as 500 and 12, respectively. The feature importance was determined using the percentage increase in mean squared error (%IncMSE). Since the total number of features is not large and all features play a crucial role in the random forest model, the adopted features are not subjected to further selection in the proposed method.

6.3.3 *Accuracy assessment*

The root mean square error (RMSE), relative RMSE (rRMSE), and coefficient of determination (R^2) were adopted to assess the differences between the observed and predicted terrain elevations of the proposed method quantitatively (Eq. (6.1)):

$$\begin{cases} \text{RMSE} = \sqrt{\dfrac{1}{n}\displaystyle\sum_{i=1}^{n}(y_i - \hat{y}_l)^2} \\ \text{rRMSE} = \dfrac{\text{RMSE}}{\bar{y}} \\ R = \frac{\sum_{i=1}^{n}(y_i-\bar{y}_i)(\hat{y}_l-\bar{\hat{y}}_l)}{\sqrt{\sum_{i=1}^{n}(y_i-\bar{y}_i)^2\sum_{i=1}^{n}(\hat{y}_l-\bar{\hat{y}}_l)^2}} \end{cases} \tag{6.1}$$

where $\hat{y}_l$ and y_i represent the predicted and observed values, respectively, and $\bar{y}$ is the mean of y_i.

6.4 Results

6.4.1 *Elevation mapping*

The predicted terrain elevations by the proposed method for the land of the study site and its surrounding shallow water areas were mapped at 10 m resolution (Figure 6.4). In general, the predicted elevations showed that the entire study site is relatively flat with a slight elevation change, expectedly corresponding with the actual terrain. The predicted terrain elevations range from −3.30 m to 4.42 m, with a mean elevation of 1.89 m. Most predicted elevations fall between 2 and 4 m. Moreover, no abnormal predicted elevations that deviated considerably from the actual elevations were found. In particular, the predicted elevations can well describe the terrain features and details. For example, the predicted elevations of the wetlands/water bodies were lower than those of the land. Low elevations were found in most areas of the study site, whereas high elevations were observed in the urban or forested regions. Moreover, we compared the predicted terrain elevations and the corresponding images of six local regions obtained from Google Earth. ICESat-2 signals can directly reach the ground in regions without dense coverage. Thus, the obtained heights of ATL03 at these regions exhibit high accuracy. As such, the predicted elevations in these regions, e.g., bare land, cropland, intertidal zone, and wetland in Figures 6.4(a), 6.4(c), 6.4(d), and 6.4(f), respectively, are relatively accurate. However, most ICESat-2 signals in regions covered by dense trees and buildings fell on the surface of nonground targets, thereby increasing the heights of ATL03 and the predicted elevations (e.g., Figures 6.4(b) and 6.4(e)). The wavelength of ICESat-2 was 532 nm, indicating good penetration performance even in turbid water. The predicted elevations at the water bodies (e.g., inland lakes/rivers, tidal creeks, and shallow seas in Figures 6.4(a), 6.4(d), and 6.4(e)) are visually lower than those in the surrounding areas.

The %IncMSE of the 35 selected features is shown in Figure 6.5. The B5 band (red edge) and MNDWI from Sentinel-2 made the greatest contribution, whereas the mean and dissimilarity of the VV polarization coefficient from Sentinel-1 had the smallest contribution.

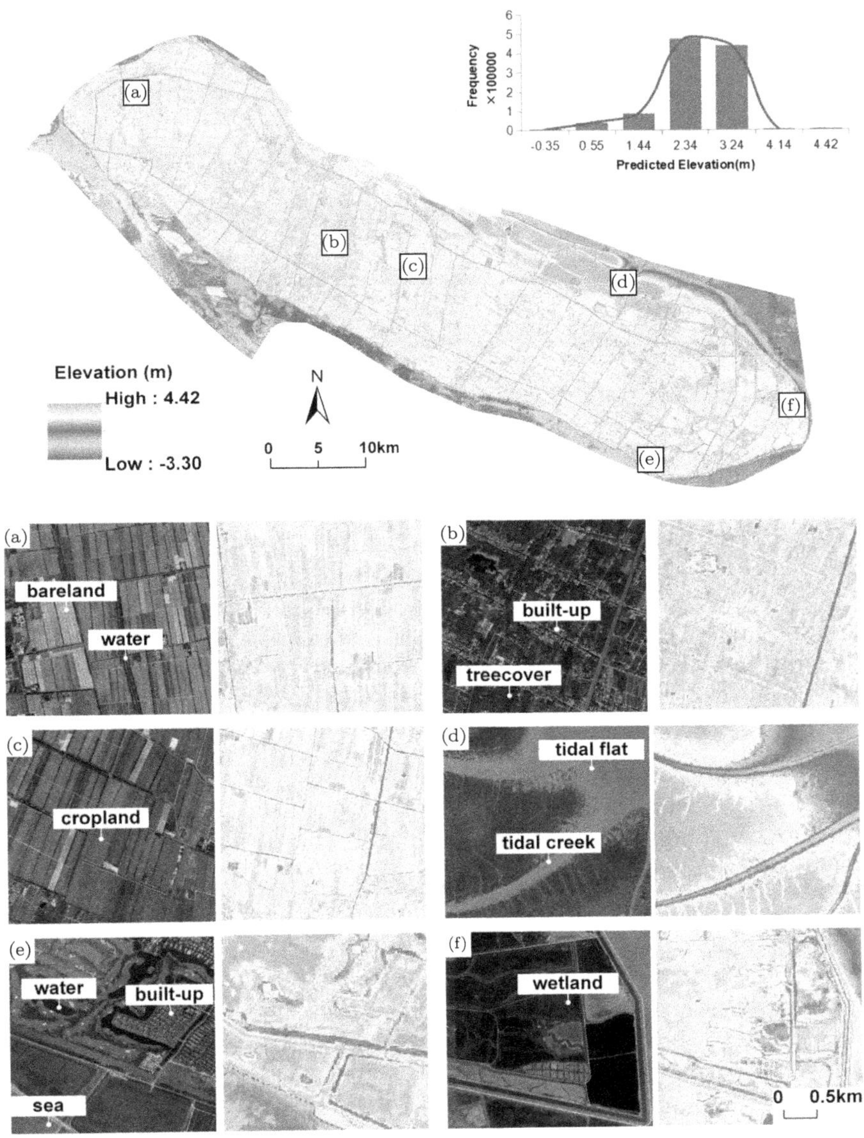

Figure 6.4. Predicted terrain elevations at 10 m resolution by the proposed method for the study site and the surrounding shallow water areas: (a)–(f) show the comparisons between the predicted terrain elevations and the corresponding images of six local regions in 2021 from Google Earth.

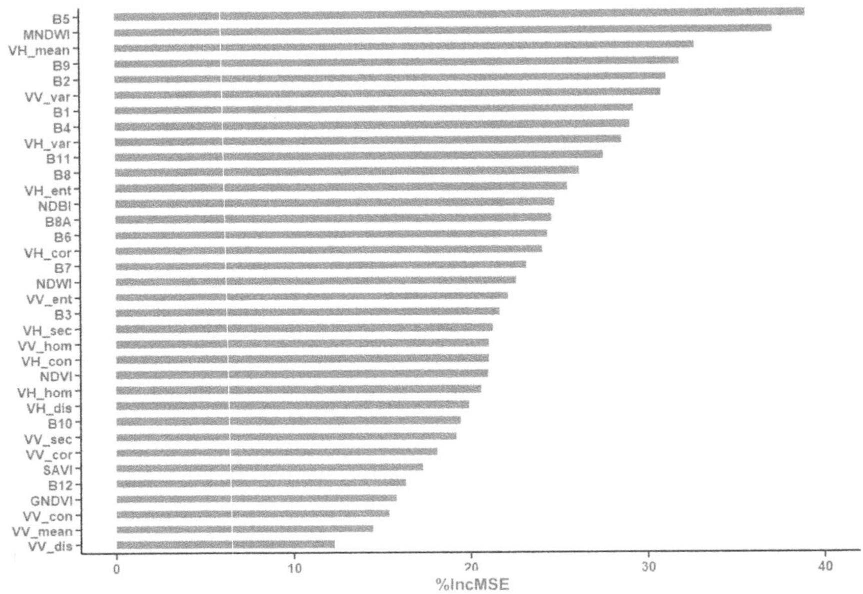

Figure 6.5. %IncMSE for different features in the random forest model, where VV/VH _mean, _hom, _var, _con, _ent, _dis, _cor, and _sec represent the mean, homogeneity, variance, contrast, entropy, dissimilarity, correlation, and second moment values of VV/VH polarization coefficients, respectively.

This result may be closely related to the land covers and landscapes of the study site. Chongming Island has a forest coverage rate of up to 25%. Chemical and physical parameters of vegetation can be effectively detected by the red edge. It is also an indicator band for depicting plant health status and pigment. Additionally, Chongming Island is surrounded by the sea and has highly developed inland water systems. Numerous rivers, lakes, ditches, canals, and harbors are distributed across the island. MNDWI can effectively identify and extract water bodies. Therefore, the B5 band and MNDWI play a key role in the random forest model. By contrast, VV polarization provides satisfactory performance in distinguishing targets with significant vertical features (e.g., tall buildings and high mountains). Chongming Island is an ecological island with flat terrain. No tall buildings or mountains exist on the island. As such, the contribution of the VV polarization features is smaller than that of other features in the random forest model.

6.4.2 *Accuracy evaluation*

The ALS ground data were interpolated to a 10 m resolution raster and converted to point elements using the inverse distance weighting method to validate the accuracy of the predicted elevations by the proposed method. The predicted terrain elevations generated by the proposed method were matched with the ALS ground points using the Extraction tool in the Spatial Analyst module of ArcGIS to conduct accuracy validation. A total of 15,377 and 23,257 pixels were obtained for the two validation regions in the Xitan and Dongtan Wetlands, respectively. The elevations measured by ALS were taken as the observed values, and the elevations estimated by the proposed method were considered the predicted values. The scatter plots of the observed and predicted elevations at each pixel are shown in Figure 6.6. High accuracy in elevation prediction was achieved in the two validation regions—Xitan Wetland (R: 0.78; RMSE: 0.59 m; rRMSE: 0.27) and Dongtan Wetland (R: 0.71; RMSE: 0.34 m; rRMSE: 0.14). The ranges of the observed elevations were 0.74–4.59 m and 0.63–2.64 m, and those of the predicted elevations were −0.52–3.17 m and 0.41–3.56 m for the two validation regions. Although the CSF algorithm and manual filtering were jointly used to ensure satisfactory acquisition of ground points from the ALS data, some ambiguous nonground points in complex scenes may still be mixed with the ground points. The unremoved nonground points may impact the accuracy evaluation of the predicted elevations.

6.4.3 *Elevation analysis for different land covers*

The study site's land cover types—tree cover, bare land, cropland, low vegetation, wetlands, and built-up areas—were individually extracted from ESA WorldCover 2020 v100 data. Then, each extracted land cover was masked by the predicted terrain elevations to obtain land cover vector data with the predicted elevation. Accordingly, the ATL08 and GEDI L2A raster data were converted to vector data. Finally, the ArcGIS "vector intersection" processing of the ATL08 and GEDI L2A data with the land cover vector data was conducted to obtain 5957 ATL08 photon points and 19,892 GEDI L2A grids that cover the study site for different land covers, where the points and grids on the sea were removed (Figure 6.7).

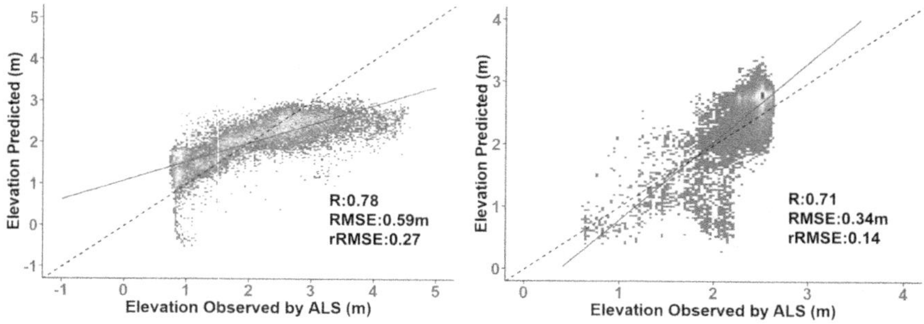

Figure 6.6. Correlation between the ALS-observed elevations and the predicted elevations of the proposed method. The dot color represents the point density, gradually increasing from blue to red: (a) validation region in the Xitan Wetland (number of points/pixels: 15,377); (b) validation region in the Dongtan Wetland (number of points/pixels: 23,257).

In particular, 2973, 401, 871, 774, 799, and 139 ATL08 photon points were allocated for cropland, wetland, tree cover, built-up, bare land, and low vegetation, whereas 10044, 1067, 4771, 2048, 1529, and 433 GEDI L2A grids were obtained for these land covers. The predicted elevations varied across different land cover types. Most negative elevations were observed in the wetlands or water bodies below mean sea level/geoid.

The predicted elevations were individually compared with the heights of ATL08 photon points and GEDI L2A grids for each land cover, considering a global perspective (Table 6.2). The results showed that the predicted elevations exhibited high consistency with the ATL08 h_te_bestfit and GEDI L2A elev_lowestmode. The RMSEs of ATL08 for cropland, wetland, tree cover, built-up, bare land, and low vegetation were 1.31, 0.72, 1.70, 1.42, 1.17, and 1.21 m, whereas the RMSEs of GEDI L2A were 0.88, 0.57, 1.21, 2.19, 1.43, and 0.80 m, respectively. The mean RMSEs were 1.26 and 1.18 m, indicating that the predicted elevations aligned with the actuality at different natural landscapes throughout the study site and that the proposed method can be effectively applied to different land covers. Compared with the predicted elevations, ATL08 h_te_bestfit and GEDI L2A elev_lowestmode data showed no significant differences across different land covers, possibly due to their equivalent accuracy. Compared with ATL08 and GEDI L2A, the predicted elevations exhibited large deviations for tree cover and built-ups. This phenomenon may be

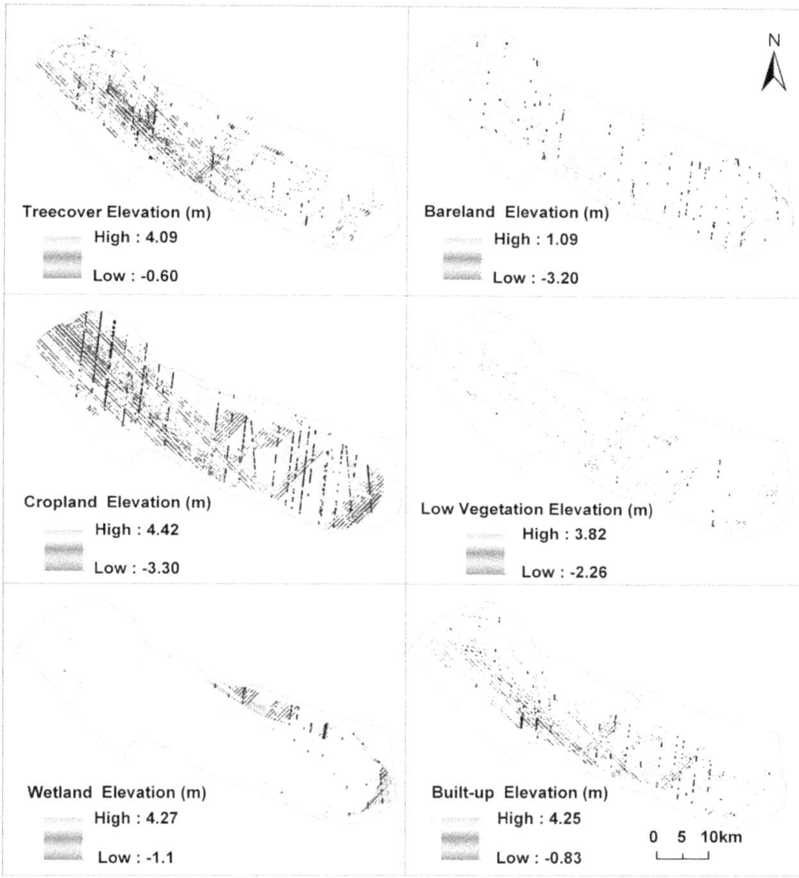

Figure 6.7. Predicted elevations under different land covers: tree cover, bare land, cropland, low vegetation, wetland, and built-up. The black and red points are the footprints of the ICESat-2 ATL08 photon points and the GEDI L2A product, respectively.

due to the dense distribution of trees and buildings and the poor penetration of laser signals at these coverages, thereby diminishing the quality of the ATL03, ATL08, and GEDI L2A original data and reducing the proposed method's prediction accuracy in these regions.

The ATL08 photon heights and GEDI L2A were compared with the global performance of the predicted elevations, rather than a

Table 6.2. Comparisons between the predicted elevations and those of ATL08 and GEDI L2A.

Reference data	Land cover type	Photon points/ grid number	R	RMSE (m)	rRMSE
ATL08	Cropland	2973	0.03	1.31	1.02
	Wetland	401	0.24	0.72	0.29
	Tree cover	871	0.09	1.70	1.68
	Built-up	774	0.27	1.42	0.97
	Bare land	799	0.27	1.17	0.78
	Low vegetation	139	0.29	1.21	0.73
GEDI L2A	Cropland	10044	0.21	0.88	0.45
	Wetland	1067	0.37	0.57	0.22
	Tree cover	4771	0.02	1.21	0.57
	Built-up	2048	0.08	2.19	0.95
	Bare land	1529	0.22	1.43	0.64
	Low vegetation	433	0.19	0.80	0.37

reference, to evaluate accuracy. By contrast, the ALS-derived elevations, with very high accuracy, served as ground truth data to assess the proposed method's performance.

6.5 Discussion

6.5.1 *Accuracy evaluation and analysis*

Compared with the ALS-derived elevations, the mean RMSE is 0.47 m for the predicted elevations in the two validation regions—Xitan and Dongtan Wetlands. This result indicates that the ATL03 data from ICESat-2 can be used for fine-resolution and high-accuracy elevation prediction with the aid of Sentinel-1/Sentinel-2 data. However, the predicted terrain elevations are generally overestimated. The possible reasons are as follows. (1) The ALS data were collected in 2020 and 2023, whereas the ICESat-2 data were collected in 2021–2022. The local elevations of the validation regions were subject to slight changes during those periods. (2) Temporal inconsistency/nonsynchronicity exists between the independent (Sentinel-1/Sentinel-2 imagery variables) and the dependent (ICESat-2 photon heights) variables of the random forest model. The annual mean images from Sentinel-1/Sentinel-2, acquired in 2021–2022, were used, whereas the ICESat-2 data were acquired at different times during this period. (3) The denoised ATL03 data still have photons of forest canopies or building surfaces, leading to an overvaluation in the seed elevations and an overestimation by the proposed model. Additionally, the accuracy of the predicted elevations in Xitan Wetland shows a relatively weaker performance than that in Dongtan Wetland. The possible reason is that the government has been carrying out extensive engineering renovations in the validation region of Xitan Wetland since 2020 and that the land cover in this tidal flat (e.g., buildings, shrubs, bare lands, trees, grasslands, and water bodies) is relatively rich. Due to the strict flight restrictions in place at the study site and the high cost of ALS data collection, only two ALS validation regions are investigated in this study.

The proposed method can predict elevations for land and water bodies (e.g., inland lakes, tidal creeks, and shallow seas). However, the seawater at the study site is extremely turbid, leading to poor penetration performance of the ICESat-2 signals. The maximum derived depth of the water body is −3.3 m and the average depth is approximately −1 m. As expected, the elevation prediction accuracy by the proposed method may not be satisfactory in the regions covered by water. In this study, the proposed method's accuracy in the water regions is not discussed due to the unavailability of *in situ* observed data. In future work, this topic can be further investigated with the assistance of the observed multibeam sonar data.

The study site is an alluvial island, and the topography is relatively flat. Theoretically, the proposed method can be applied to other areas with significant elevation differences. However, the random forest model should be reconstructed, and the importance of the selected features may change when the proposed method is applied in other areas with different land cover types and elevation features. In the proposed method, Sentinel-1 and Sentinel-2 can be substituted with other similar remotely sensed data. Additionally, remotely sensed data, together with Sentinel-1 and Sentinel-2, can serve as independent variables to improve the accuracy of terrain elevation estimation.

6.5.2 *Accuracy comparisons for different land covers*

Several previous studies have attempted to estimate elevations/topography in regions with a single landscape by using ICESat-2 data, e.g., forest (Li *et al.*, 2023), shallow water (Xie *et al.*, 2023), and polar regions (Farrell *et al.*, 2020). Few studies have utilized ICESat-2 data to predict elevations for a spatially continuous, large region with multiple land cover types. The proposed method can achieve good prediction performance for any landscape because different land covers exhibit varied elevations and corresponding polarization and spectral characteristics. Given the field data limitations, only ALS data in two local regions were used to validate the predicted elevations. The ATL08 and GEDI data were compared globally with the predicted elevations under different land covers. It should be noted that ATL08 or GEDI data can also be integrated with Sentinel-1 and Sentinel-2 for terrain elevation estimation. However, ICESat-2 ATL03 was used in this study instead of ATL08/GEDI, as the ATL08/GEDI ground photons are insufficient for mapping terrain elevations at high resolution and accuracy. For example, only 5,957 ATL08 photon points can be used, while 25,687 ATL03 photon points are contained for the study site. ICESat-2 and GEDI are expected to have geolocation errors and elevation measurement errors. These error sources and budgets were ignored in the study and should be corrected in future work to further improve accuracy.

6.6 Conclusion

An innovative methodology is developed to map terrain elevations by coupling the ICESat-2, Sentinel-1, and Sentinel-2 satellite remotely sensed data. The following major conclusions can be drawn from this study:

(1) ICESat-2 possesses the capability for spatially continuous, large-scale terrain elevation estimation under multiple land covers, with the aid of Sentinel-1 and Sentinel-2. Decimeter-level vertical accuracy and 10 m spatial/horizontal resolution can be achieved for the estimated terrain elevations.
(2) The estimated terrain elevations from ICESat-2 ATL03 show good consistency with those derived from ATL08 and GEDI L2A,

indicating equivalent accuracy performance for the two advanced spaceborne laser sensors and their corresponding products.

Compared with existing studies, which focus on terrain elevation estimation for single land covers, the proposed method exhibits high accuracy and satisfactory robustness performance across different land covers. Thus, it provides a very promising and supplementary solution for large-area, fine-resolution elevation mapping. The effectiveness and time-series performance across different land covers and landforms (e.g., hills and mountains) should be comprehensively investigated further in future work. With abundant global data, the proposed method may be applied to elevation estimation even at global scales. Only random forest is used in this study. Advanced machine learning models or deep learning can be comprehensively investigated in future work.

Chapter 7

Dissolved Oxygen Classification and Regression in the Taihu Lake Basin with Bidirectional Long Short-Term Memory Network based on Physical and Chemical Prior Knowledge

Dissolved oxygen (DO) is a key parameter for monitoring water quality, as it provides important guidance for aquaculture and environmental warnings. However, previous studies have overlooked the physicochemical relationships between DO and other parameters in the process of simulating DO, resulting in simulated values that do not correspond to actual physical laws. Additionally, the opacity of deep learning models limits their applicability. Therefore, this study proposes a model for predicting and classifying DO concentration in the Taihu Lake Basin based on a prior knowledge-guided bidirectional long short-term memory network (PKBiLSTM), aiming to classify and predict DO concentrations at 16 hydrological stations in the basin. Furthermore, Shapley additive explanations (SHAP) are used to analyze the impact of internal parameters on DO. To further validate the effectiveness and competitiveness of the PKBiLSTM model, we compared it with nine other commonly used machine learning (ML) algorithms, including convolutional neural networks (CNNs) and BiLSTM. The results show that the overall performance of the PKBiLSTM model is superior to that of the other models across nine datasets. Specifically, the average Kling–Gupta efficiency coefficient (KGE) of the PKBiLSTM model is 0.926, which is 2.48% and 2.38% higher than the CNN and GRU models, respectively. Meanwhile, on the B8 and B15 test sets, the F1 scores of the PKBiLSTM model are 0.987 and 0.897, respectively, which are 3.44% and 1.45% higher than the random forest (RF) model. The experiments show that an increase in pH corresponds to an increase in DO concentration when the pH value is between 6.5 and 10, consistent with established knowledge in the field. Additionally, the model achieves the highest accuracy when the input variables are water temperature, pH, conductivity, and total nitrogen (TN). This method can provide a reference and theoretical basis for the scientific management and monitoring of water resources.

7.1 Introduction

Rivers play a crucial role in providing essential water resources for sustaining human life (Pandhiani *et al.*, 2020; Asadollah *et al.*, 2021). However, with the intensification of industrialization and urbanization, water pollution has progressively escalated, exacerbating ecological and environmental challenges in various water bodies worldwide, especially in developing countries (Chen *et al.*, 2023). Among the numerous water quality indices, DO is one of the most crucial parameters, as it directly influences the health of aquatic ecosystems and their biological diversity (Banerjee *et al.*, 2019; Banks *et al.*, 2012; Kannel *et al.*, 2007; Liu *et al.*, 2021). DO concentration in water bodies varies greatly as it depends on: (i) environmental parameters (mainly temperature and precipitation); (ii) biological indicators such as the abundance of algae and fish; and (iii) the input of wastewater derived from human activities. Hence, the development of modeling techniques capable of providing accurate information on DO levels is vital, as they can provide valuable insights and theoretical support for local environmental governance (Albuquerque *et al.*, 2021; Schmidtko *et al.*, 2017).

DO prediction can generally be categorized into two types: process-driven and data-driven approaches (Qiu *et al.*, 2020; Kim *et al.*, 2022; Yu *et al.*, 2022; Nguyen *et al.*, 2018). Process-driven approaches, such as Row Column AESOP (Wang *et al.*, 2017), focus on the numerical representation of the physical, chemical, and biological processes of water bodies and are mainly based on the law of mass conservation to simulate DO variation rules (Gelda *et al.*, 2019; Yan *et al.*, 2021). However, implementing such approaches faces significant challenges arising from the intricate interactions between water systems and other intertwined systems (Zhu *et al.*, 2023; Wang *et al.*, 2022), and they require a substantial amount of computing power. Furthermore, the escalating demand for computing power required by real-time simulations presents additional obstacles, complicating the implementation of such methodologies (Fu *et al.*, 2022).

In contrast to process-driven models, data-driven methods do not require considering the physical and biochemical changes between DO and other factors (Bedri *et al.*, 2014). Instead, they solely rely on historical data to simulate DO concentrations. Data-driven methods can be broadly divided into two categories: (i) traditional statistics-based models; and (ii) ML-based models (Alexakis *et al.*, 2013). Traditional statistics-based models are well-regarded for their simplistic structural design and ease of computation (Yang *et al.*, 2023). Commonly used statistics-based DO prediction models include the hidden Markov model (Li *et al.*, 2022), the gray model (Li *et al.*, 2018), and polynomial chaos expansions (PCE) (Keshtegar *et al.*, 2019). However, these techniques usually require the assumption of a linear correlation between variables, while the computational complexity increases rapidly with the number of variables, constraining the application of such models (Zheng *et al.*, 2023).

On the other hand, ML-based models are capable of capturing nonlinear relationships between features and analyzing influencing factors without the need to understand the underlying physical processes (Kruk, 2023). Commonly used ML-based DO prediction models include the adaptive neuro-fuzzy inference system (ANFIS) (Najah *et al.*, 2014; Kisi *et al.*, 2020), least squares support vector machines (Song *et al.*, 2021), regularized extreme learning machines (RELMs) (Cao *et al.*, 2019), and random forest (RF) (Valera *et al.*, 2020). Dehghani *et al.* (2022) used four optimization algorithms, including the innovative Gunner's Algorithm, to optimize the hyperparameters of a support vector regression (SVR) and predict DO concentrations in the Cumberland River in the southern US.

Over the past decade, deep learning, represented by neural networks, has attracted widespread attention due to its excellent performance in handling big data. Hence, various deep learning techniques such as long short-term memory (LSTM) (Wu and Wang 2022; Zhi *et al.*, 2021) and artificial neural networks (ANNs) (Mitrović *et al.*, 2019; Zhu and Heddam 2020; Imani *et al.*, 2021) are widely used in DO prediction. For example, Barzegar *et al.* (2020) constructed LSTM, CNN, and an ensemble of CNN–LSTM models to predict DO concentrations. The results indicated that the CNN–LSTM model outperformed all other models in terms of accuracy. Bi *et al.* (2023) developed a model, namely SVABEG, to predict DO concentrations. Guo *et al.* (2022) took the DO concentrations of sea bass breeding as the research object. Initially, principal component analysis (PCA) was adopted to select relevant external variables and improve calculation efficiency. Subsequently, a gated recurrent unit (GRU) optimized by the pathfinder optimization algorithm was used to predict the DO concentrations.

Over the last decade, ML techniques have constantly been developed for the prediction of DO concentration, yielding significant outcomes. However, there are still several challenges that need to be addressed carefully.

(1) While deep learning models have demonstrated remarkable abilities in simulating DO, they primarily model DO changes based on data features and ignore the physical and chemical relationships between DO concentrations and other factors within the water body. Hence, there is a potential risk that the predicted results diverge from the actual physical rules (Karpatne *et al.*, 2017; Xu and Liang, 2021; Varadharajan *et al.*, 2022).
(2) Although the accuracy of DO concentration predictions can be greatly increased through the application of ML models, as model complexity rises, interpretability frequently suffers as well. As a result, it becomes harder to pinpoint the precise causes of the fluctuations in DO (Lee *et al.*, 2022; Zheng *et al.*, 2023).
(3) Considering the complexity and nonlinearity of DO concentrations, incorporating all external factors as input to the neural network would increase computational complexity. Moreover, it may also lead to issues such as reduced prediction accuracy due to interactions among multiple variables.

This study focuses on the DO concentration at 16 hydrological stations in the Taihu Lake Basin. Considering that BiLSTM can fully consider past and future information when processing time-series data, as well as the physical and chemical relationships between DO concentration and other factors in the water body, this research utilizes it as prior knowledge to modify the structure of the BiLSTM model and proposes the PKBiLSTM model. Meanwhile, we compare the accuracy and stability of PKBiLSTM with nine other models, such as CatBoost, which have been recently reported for DO concentration prediction, to validate the effectiveness of PKBiLSTM. Furthermore, due to the black-box nature of deep learning, which greatly limits the applicability of the model, SHAP is adopted to analyze the impact of each input variable on DO concentration from both a specific time point and a global perspective. The innovations of this research are listed as follows:

(1) Considering the physical relationship between DO and water temperature, as well as the potential chemical reaction between DO and COD_{mn}, this study incorporates them as prior knowledge to modify the structure of the BiLSTM to constrain the output.
(2) Recognizing the fact that the credibility of traditional ML models is often compromised due to the inherent lack of transparency, the study introduces SHAP to evaluate the contribution of each input variable to the DO values. SHAP provides a systematic and interpretable approach to understanding the impact of individual features on the target variable, thus enhancing the transparency and trustworthiness of the model.
(3) To further determine the optimal input variables for DO concentration prediction, this study used the prediction error evaluation method to analyze the prediction performance of different combinations of input variables.

The remainder of this study is organized as follows. Section 7.2 primarily introduces the research locations and data sources. Section 7.3 focuses on the design of PKBiLSTM and presents the overall framework of the study. Section 7.4 presents the classification and regression results of the PKBiLSTM and other models on the Taihu Lake Basin test set. Section 7.5 focuses on analyzing the classification and regression results of each model at different sites and the contribution of input variables to DO concentrations. Finally, Section 7.6 provides the conclusions and outlook of the study.

7.2 Study Site and Data Source

7.2.1 *Study site*

The Taihu Lake Basin is located in the southern portion of the Yangtze River Delta, with the Yangtze River to the north, the East China Sea to the east, the Qiantang River to the south, and the Tianmu and Maoshan mountain climbing areas to the west, encompassing a watershed area of about 36,900 km^2 44. The basin is an important region in China with high population density and intense economic activity, and the gross domestic product (GDP) of the basin accounts for about 10.2% of the national GDP.

7.2.2 *Data source and analysis*

The primary source of data used in this research is the China National Surface Water Quality Real-time Monitoring Data Release System (https://szzdjc.cnemc.cn:8070/GJZ/Business/Publish/Main.html). The data collected from the monitoring stations have a four-hour sampling frequency and cover the period from November 1, 2020, to February 28, 2023. Figure 7.1 presents information on the specific location of each monitoring site.

The names and abbreviations of various water quality monitoring stations are shown in Table 7.1.

The dataset includes key water quality parameters such as total nitrogen (TN) and permanganate index (COD_{mn}) (Table 7.2).

Subsequently, SPSS26 software was employed to perform descriptive statistical analysis on the dataset. Take the B3 dataset as an example. Table 7.3 displays the descriptive statistical analysis results for the B3 dataset.

Standard deviation is a statistical measure that quantifies the dispersion of data points relative to the mean. From Table 7.3, the standard deviation of the DO sequence is 2.234, indicating a relatively large dispersion of data points around the mean. Skewness is a statistical measure that describes the degree of asymmetry in the distribution of data. More specifically, a skewness of 0.149 indicates that the DO sequence is slightly skewed toward positive skewness, meaning there is a greater concentration of data points in the right tail of the distribution. As a result, applying conventional statistical

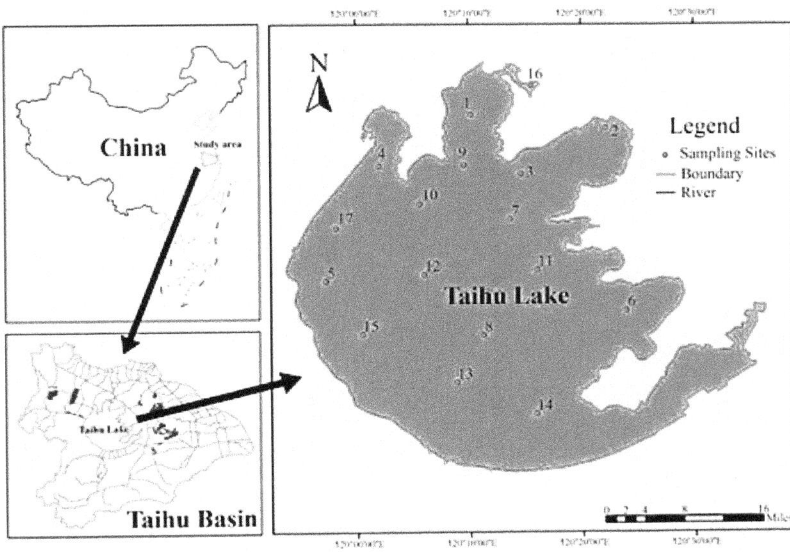

Figure 7.1. The information on the specific location of each monitoring site.

Table 7.1. The names and abbreviations of various water quality monitoring stations.

Name	Abbreviation
Baiyukou	B1
Baofeng Village Estuary	B2
Caohai Center	B3
Chuanfang Wushe Qiaotou (Station One)	B4
Daguan River Estuary	B5
Tuluo Village Inlet, Dayu Township	B6
Broken Bridge	B7
Guanyin Mountain East	B8
Guanyin Mountain West	B9
Wuli Lake Central	B10
Huilong Village	B11
Jiangwei Lower Lock	B12
Luojia Ying	B13
Wang Bridge	B14
Xinhe Village E	B15
Yanjia Village Bridge	B16

Table 7.2. Details of each water quality parameter.

Index	Variable	Unit
A1	Water temperature	°C
A2	pH	/
A3	Permanganate index (COD_{mn})	mg/L
A4	Ammonia nitrogen (AN)	mg/L
A5	Total phosphorus (TP)	mg/L
A6	Total nitrogen (TN)	mg/L
A7	Electric conductivity	$\mu\frac{s}{cm}$
A8	Turbidity	NTU
A9	Chlorophyll	mg/m^3
A10	Algal density	mg/m^3

Table 7.3. The descriptive statistical analysis of the B3 dataset.

		N	Min	Max	Mean	Std	Skewness	Kurtosis
B1	A1	4202	8.500	28.730	17.184	4.566	0.066	−1.254
	A2	4202	4.024	10.190	8.573	0.472	−1.329	10.665
	A3	4202	1.405	21.020	6.818	1.405	2.544	20.423
	A4	4202	0.025	3.210	0.204	0.237	2.182	13.381
	A5	4202	0.005	0.156	0.062	0.019	0.848	0.889
	A6	4202	0.545	9.496	1.572	0.433	2.552	35.125
	A7	4202	344.100	511.650	427.459	34.943	−0.262	−0.550
	A8	4202	4.957	295.000	29.138	17.567	2.947	23.697
	A9	4202	0.000	0.138	0.009	0.014	3.620	15.486
	A10	4202	27698	267000000	24933663	16825513	1.461	10.776
	A11	4202	1.961	15.857	7.678	1.503	0.407	2.092
B2	A1	4210	7.300	28.010	18.058	3.709	−0.153	−0.827
	A2	4210	7.160	9.620	8.282	0.435	0.378	−0.559
	A3	4210	0.250	13.538	2.859	1.223	3.562	18.423
	A4	4210	0.025	6.937	0.198	0.452	5.365	42.914
	A5	4210	0.005	1.076	0.104	0.076	3.900	25.739
	A6	4210	0.419	19.726	4.261	1.236	2.589	24.598
	A7	4210	1.450	814.190	587.638	62.127	−0.951	5.809
	A8	4210	3.860	1355.990	24.681	73.129	9.398	114.236
	A11	4210	1.460	28.580	8.018	4.449	1.387	1.272
B3	A1	4155	8.810	27.840	17.745	4.649	0.055	−1.298
	A2	4155	7.452	9.555	8.519	0.295	−0.151	0.281
	A3	4155	0.964	22.343	4.221	1.711	1.450	4.600
	A4	4155	0.025	0.556	0.064	0.066	2.376	5.983
	A5	4155	0.005	0.483	0.061	0.025	3.183	37.830
	A6	4155	0.082	7.361	3.631	0.927	−0.170	−0.418
	A7	4155	334.370	511.400	415.935	31.830	0.371	−0.240
	A8	4155	2.900	214.180	21.435	17.128	4.139	29.203
	A9	4155	0.000	0.122	0.017	0.013	2.721	11.432
	A10	4155	195124	254000000	31456893.54	30463652.75	1.841	5.290
	A11	4155	0.883	20.254	9.881	2.234	0.149	0.657

methods to simulate DO concentrations may achieve unsatisfactory results.

7.3 Materials and Methods

7.3.1 *Prior knowledge constraints bidirectional long short-term memory network*

7.3.1.1 *Bidirectional long short-term memory network*

An LSTM network is a classical deep learning model. Unlike traditional recurrent neural networks (RNNs), LSTMs incorporate forget, input, and output gates, thus enabling them to effectively overcome the problem of gradient vanishing and explosion when dealing with long time sequences (Hochreiter and Schmidhuber, 1997). However, LSTMs can only predict information for the next time step based on that of the previous time step. Compared to the LSTM model, the BiLSTM model can fully capture dependency information in both directions of the data, resulting in more effective feature extraction of the input data. The final output of the model is obtained by combining the outputs calculated by the forward and backward LSTM layers after processing the input sequence (Xiong *et al.*, 2023; Graves and Schmidhuber, 2005; Geng *et al.*, 2023; Zhang *et al.*, 2023) (Eq. (7.1)):

$$y_t = \sigma(W_y \cdot [\overrightarrow{h_t}, \overleftarrow{h_t}] + b_y) \tag{7.1}$$

$\overrightarrow{h_t}$ and $\overleftarrow{h_t}$ are the forward and backward hidden layer states at moment t, respectively. y_t is the output at moment t. W_y and b_y are the weight matrix and bias terms, respectively.

7.3.1.2 *Prior knowledge constraints*

ML models have been extensively applied to DO prediction. However, these models only consider DO changes from the perspective of data features, ignoring the physical and chemical relationships between DO and other parameters within water bodies. Meanwhile, most of the prior studies have typically adopted root mean square error (RMSE) as a loss function for neural networks to quantify the difference between predicted and actual values (Bommidi *et al.*, 2023; Li *et al.*, 2024; Chen *et al.*, 2022), which can easily lead to discrepancies between the predicted results and the actual physical patterns. Therefore, considering that DO in rivers undergoes physicochemical changes related to other parameters such as water temperature and COD_{mn}, this study incorporates it as prior knowledge to modify the structure of the BiLSTM model to align the anticipated outcomes with the observed physical patterns.

(1) Physical constraint

Water temperature has an extremely crucial effect on DO concentration (Jane *et al.*, 2021; Nong *et al.*, 2021). Specifically, heterotrophic respiration grows at a faster rate than primary production as water temperature increases, thus accelerating the consumption of DO concentrations (Yvon *et al.*, 2010). Meanwhile, the saturation of DO decreases significantly with increasing water temperature (Thomann and Mueller, 1987; Nong *et al.*, 2023; Zhou and Zhang, 2023). The relationship between DO and temperature under the same pressure is displayed in Eq. (7.2):

$$\ln c_{sf} = -139.34411 + \frac{1.575701 \times 10^5}{T} - \frac{6.642308 \times 10^7}{T^2} + \frac{1.2438 \times 10^{10}}{T^3} - \frac{8.621949 \times 10^{11}}{T^4} \tag{7.2}$$

where T represents temperature, and c_{sf} denotes the saturation of DO in the water at one atmosphere pressure (mg/L).

From the above equation, it can be observed that under the same atmospheric pressure, DO saturation exhibits a decreasing trend with increasing temperature, as shown in Eq. (7.3):

$$\left\{ \begin{array}{l} \widehat{y}^k_{t+p} - y^k_t > 0, T^k_{t+p} - T^k_t < 0 \\ \widehat{y}^k_{t+p} - y^k_t < 0, T^k_{t+p} - T^k_t > 0 \end{array} \right\} \tag{7.3}$$

where y^k_t represents the observed DO concentrations value at time t at location k. $\widehat{y}^k_{t+p}$ represents the simulated DO value at time $t+p$ at location k. Moreover, considering the non-negativity of DO itself, the loss function based on physical constraint is modified in Eq. (7.4):

$$\text{Loss}_p = \frac{1}{N_k N_L} \sum_{i=1}^{N_k} \sum_{j=1}^{N_L} \text{ReLU}[(\widehat{y}^k_{t+p} - y^k_t)(T^k_{t+p} - T^k_t)] \tag{7.4}$$

N_k represents the number of individual hydrological monitoring stations. N_L represents the length of the dataset.

(2) Chemical constraints

Furthermore, in an aqueous environment, COD_{mn} undergoes redox reactions with DO. Therefore, as the concentration of COD_{mn} increases, the concentration of DO decreases. The loss function based on chemical constraints is modified in Eq. (7.5):

$$\mathrm{Loss}_c = \frac{1}{N_k N_L} \sum_{i=1}^{N_k} \sum_{j=1}^{N_L} \mathrm{ReLU}[(\widehat{y}_{t+p}^k - y_t^k)(T_{t+p}^k - T_t^k)] \tag{7.5}$$

The total loss function Loss_T is calculated as shown in Eq. (7.6):

$$\mathrm{Loss}_T = \mathrm{RMSE} + \gamma \mathrm{Loss}_p + \delta \mathrm{Loss}_c \tag{7.6}$$

γ and δ represent the weight coefficients, respectively.

7.3.2 *Shapley additive explanations*

Considering the fact that the inherent black-box nature of traditional ML algorithms reduces the models' trustworthiness and applicability, SHAP is incorporated in this work. SHAP is a powerful tool for quantifying the contribution of each individual feature toward final model predictions. It has been extensively applied in the fields of energy consumption prediction (Amiri *et al.*, 2023), landslide prediction (Al-Najjar *et al.*, 2022), and other domains. The Shapley value is a method in cooperative game theory that allocates the total payoff fairly among members of a team based on their contributions to the overall payoff. The value is calculated as shown in Eq. (7.7):

$$\varphi_j(\mathrm{val}) = \sum_{s \in \{x_1, \ldots, x_p\} / \{x_j\}} \frac{|s|!(p - |s| - 1)!}{P!} (\mathrm{val}(s \cup \{x_j\}) - \mathrm{val}(s)) \tag{7.7}$$

where $\varphi_j(\mathrm{val})$ refers to the shapley value of feature j, i.e., the contribution to the result; s is the subset of features used in the model; x is the vector of sample feature values to be interpreted; p is the number of features; $\frac{|s|!(p-|s|-1)!}{P!}$ refers to the weights; and $\mathrm{val}(s)$ refers to the value of the model output under the combination of features s. The interpreted result of the model is as shown in Eq. (7.8):

$$g = \varphi_0 + \sum_{i=1}^{M} \varphi_j \tag{7.8}$$

where g is the explanatory model, M is the number of input features, φ_j is the shapley value of feature j, and φ_0 is a constant referring to the predicted mean of all the samples.

7.3.3 *Data normalization*

To mitigate the impact of varying units and scales of input variables on the results, minimum-maximum (min-max) normalization was employed to preprocess the data. Min-max normalization is a popular method used to rescale original data, ensuring that all variables are proportionally adjusted within a specified range. By applying this technique, the values of the variables are transformed to a standardized scale, typically between 0 and 1, without distorting their relative relationships (Chen, 2019).

7.3.4 *Evaluation metrics*

In this research, the Kling–Gupta efficiency coefficient (KGE) and four other metrics were used to comprehensively evaluate regression outcomes from multiple perspectives. Meanwhile, F1 score, precision, accuracy, recall, and the area under the curve (AUC) were adopted to assess the classification results of each model. The definition and formula of each evaluation index are displayed in Table 7.4.

7.3.5 *Study framework*

The framework of this study is presented as follows:

(1) The relevant data on the water quality of the Taihu Lake basin are gathered.
(2) Data pre-processing: Considering the potentially adverse impact of missing values on the final results of a model, the K-nearest neighbors (KNN) algorithm is employed to handle missing values (Sareen *et al.*, 2023; Park *et al.*, 2022).
(3) Data normalization: In this study, the min-max normalization method was introduced to remove the magnitude differences between different data.
(4) Splitting the dataset: The water quality dataset is split into training, validation, and test sets in the ratio of 6:2:2.
(5) Model training and testing: The normalized water temperature, pH, and other water quality data are introduced as inputs to

Table 7.4. The definition and formula of each evaluation index.

Index	Meaning	Formula
RMSE	Root mean square error	$\text{RMSE} = \sqrt{\frac{1}{N}\sum_{t=1}^{N}(y_o(t) - y_s(t))^2}$
MAE	Mean absolute error	$\text{MAE} = \frac{1}{N}\sum_{t=1}^{N}\lvert y_o(t) - y_s(t)\rvert$
NSE	Nash–Sutcliffe efficiency	$\text{NSE} = 1 - \sum_{t=1}^{N}\frac{(y_o(t)-y_s(t))^2}{(y_o(t)-\bar{y_o})^2}$
KGE	Kling–Gupta efficiency coefficient	$\text{KGE} = 1 - \sqrt{(r'-1)^2 + (\alpha-1)^2 + (\beta-1)^2}$
R^2	Coefficient of determination	$R^2 = \left(\frac{\sum_{t=1}^{N}(y_o(t)-\bar{y}_o)(y_s(t)-\bar{y}_s)}{\sqrt{\sum_{t=1}^{N}(y_o(t)-\bar{y}_o)^2}\sqrt{\sum_{t=1}^{N}(y_s(t)-\bar{y_s})^2}}\right)^2$
Accuracy	Accuracy	$\text{Accuracy} = \frac{\text{TP+TN}}{\text{TP+TN+FP+FN}} \times 100\%$
Recall	Recall	$\text{Recall} = \frac{\text{TP}}{\text{TP+FN}} \times 100\%$
Precision	Precision	$\text{Precision} = \frac{\text{TP}}{\text{TP+FP}} \times 100\%$
F1	F1 score	$F1 = \frac{2\times\text{Precision}\times\text{Recall}}{\text{Precision}+\text{Recall}}$
AUC	Area under the curve	$\int_0^1 \frac{\text{TP}}{\text{TP+FN}}\, d\frac{\text{FP}}{\text{FP+TN}}$
CM	Confusion matrix	$\begin{bmatrix}\text{TP} & \text{FP}\\ \text{FN} & \text{TN}\end{bmatrix}$

Note: y_o is the actual value. $\bar{y_o}$ and $\bar{y_s}$ are the mean of the actual value and simulation value, respectively. y_s is the predicted value, and N is the total number of data sets. α and β are the ratios of the standard deviation and mean of the simulation value to the actual value, respectively. r' is the correlation coefficient between actual and simulation values. *TP*, *TN*, *FP*, and *FN* are true positives, true negatives, false positives and false negatives, respectively. The details of *TP*, *TN*, *FP* and *FN* are displayed in Table 7.5.

the PKBiLSTM, LSTM, CatBoost, and other models, and the normalized DO is used as the output to train the models and make predictions on the test set.

(6) Model evaluation: In this study, F1, recall, NSE, KGE, and other metrics are used to assess the regression and classification performance of each model on the training, validation, and test sets of each water quality dataset.
(7) Input importance analysis: SHAP is employed to analyze the contribution of internal parameters of water bodies between different sites to DO concentrations.

The framework of this study is presented in Figure 7.2.

Table 7.5. The definition of the TP, TN, FP and FN.

Index	Definition
TP	True Positive
TN	True Negative
FP	False Positive
FN	False Negative

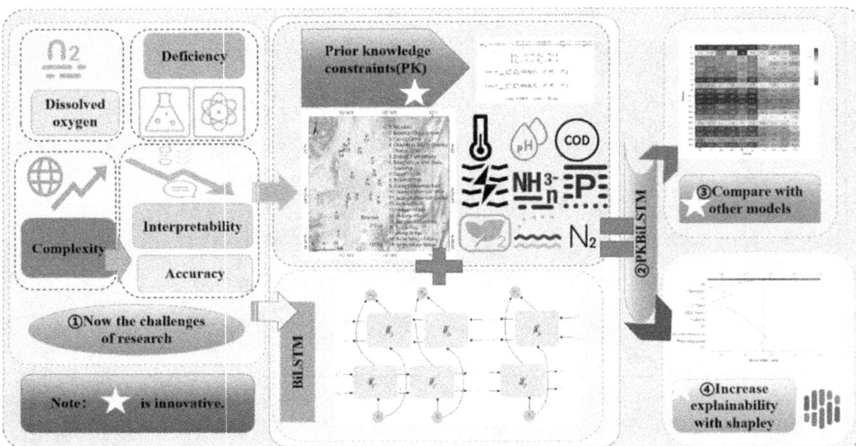

Figure 7.2. The framework of this study.

7.4 Results

7.4.1 *Model comparison*

To assess the effectiveness of the PKBiLSTM model, nine different models were used for comparison against it. The introduction of other models is presented as follows.

CatBoost: Given the iterative nature of traditional ensemble learning algorithms such as AdaBoost and gradient boosting decision trees (GBDTs), there is a common issue of overfitting when dealing with high-dimensional data. To tackle this challenge, the CatBoost algorithm introduces a gradient descent-based ordered boosting framework that improves upon GBDT by optimizing the fully symmetric decision trees more effectively (Dorogush *et al.*, 2018).

eXtreme Gradient Boosting (XgBoost): XgBoost is a gradient boosting algorithm that was developed based on the GBDT algorithm. In the framework of residual-based training, it performs k iterations of training, where each iteration utilizes a different binary tree (usually a CART) for training. This iterative process helps to address the shortcomings of the model after each training iteration and improves prediction accuracy. The final prediction is obtained by summing up the predictions generated by the k-training iterations (Chen and Guestrin, 2016; Dong *et al.*, 2023).

RF: The RF algorithm, proposed by Breiman in 2001 (Breiman, 2001), is an ML method based on decision trees. Since decision trees have intuitive structures and strong interpretability, they have been widely applied to various domains. However, decision trees are prone to overfitting, which limits their performance. In the RF algorithm, multiple decision trees are trained on random subsets of the data, known as bootstrap samples. Each decision tree is trained independently on a different subset of the data, ensuring diversity among the individual trees. Additionally, at each split in the tree, a random subset of features is considered, further promoting diversity and reducing correlation between the trees (He *et al.*, 2022; Vijh *et al.*, 2020).

GRU: GRU is a variant of LSTM that can solve the gradient explosion and gradient vanishing problems in RNNs (Yao *et al.*, 2023; Min *et al.*, 2023). Compared with LSTM, GRU has only two gates: the update gate, which controls the range and retains previous information in the current state; and the reset gate, which determines historical information and whether the current state should be turned off (Lv *et al.*, 2023). Therefore, GRU has the advantages of a simpler structure, a smaller number of parameters, and easier convergence of model training (Liu *et al.*, 2023).

The abbreviation for each model is presented in Table 7.6.

7.4.2 *Model parameters*

The parameters of each model are displayed in Table 7.7.

7.4.3 *Regression evaluation results for each model*

The water quality data from stations B1–B16 are taken as the research objects. Following the principles of ML for dataset splitting,

Table 7.6. The abbreviation of each model.

Model	Abbreviation
PKBiLSTM	M1
BiLSTM	M2
LSTM	M3
GRU	M4
Backpropagation (BP)	M5
CNN	M6
CatBoost	M7
XgBoost	M8
RF	M9
SVM	M10

Table 7.7. The parameters of each model.

Model	Parameters	Value
PKBiLSTM/ BiLSTM/ LSTM/GRU	The maximum number of iterations	200
	Batch Size	32
	Learning rate	0.003
	The number of neurons in the hidden layer	100
	The number of hidden layers	3
	Optimizer	Adam
	Loss function	RMSE
BP	The maximum number of iterations	200
CNN	The number of neurons in the hidden layer	100
CatBoost/ XgBoost	Learning rate	0.003
	Depth	6
	Number of base learners	200
RF	Number of base learners	200
SVM	Kernel function	Radial basis function
	Penalty factor	10
	Epsilon	0.005

Table 7.8. The performance of each model on the training and validation sets of the B1 and B2 datasets.

		Training Set					Validation Set				
Datasets	Model	RMSE	MAE	NSE	KGE	R^2	RMSE	MAE	NSE	KGE	R^2
B1	M1	0.390	0.258	0.930	0.903	0.966	0.700	0.516	0.858	0.823	0.931
	M2	0.398	0.260	0.927	0.909	0.964	0.708	0.525	0.855	0.809	0.931
	M3	0.445	0.289	0.909	0.875	0.956	0.707	0.527	0.855	0.842	0.929
	M4	0.388	0.252	0.931	0.913	0.966	0.684	0.501	0.865	0.833	0.935
	M5	0.401	0.259	0.926	0.910	0.963	0.687	0.501	0.864	0.827	0.936
	M6	0.388	0.252	0.931	0.909	0.966	0.690	0.506	0.862	0.816	0.936
	M7	0.371	0.243	0.936	0.897	0.970	0.613	0.456	0.891	0.848	0.948
	M8	0.397	0.260	0.927	0.904	0.964	0.715	0.530	0.852	0.795	0.932
	M9	0.398	0.260	0.927	0.909	0.964	0.708	0.525	0.855	0.809	0.931
	M10	0.430	0.280	0.915	0.883	0.959	0.745	0.555	0.839	0.783	0.926
B2	M1	1.589	1.141	0.894	0.866	0.948	1.363	0.970	0.744	0.822	0.865
	M2	1.527	1.072	0.902	0.882	0.952	1.257	0.885	0.783	0.819	0.885
	M3	2.109	1.515	0.814	0.802	0.905	1.684	1.195	0.610	0.688	0.783
	M4	1.738	1.221	0.873	0.854	0.937	1.419	0.993	0.723	0.785	0.850
	M5	2.806	2.085	0.670	0.677	0.825	2.246	1.739	0.306	0.433	0.584
	M6	1.675	1.185	0.882	0.866	0.942	1.355	0.957	0.747	0.793	0.865
	M7	2.086	1.482	0.818	0.798	0.909	1.613	1.123	0.642	0.679	0.804
	M8	1.993	1.465	0.833	0.818	0.917	1.666	1.223	0.618	0.748	0.790
	M9	2.070	1.461	0.820	0.805	0.911	1.637	1.141	0.631	0.717	0.796
	M10	2.120	1.347	0.812	0.845	0.901	1.393	0.981	0.733	0.817	0.859

we set the first 60% of the data as the training set, 20% as the validation set, and the last 20% as the test set. Table 7.8 presents the performance of each model on the training and validation sets of the B1 and B2 datasets.

Table 7.9 presents the error evaluation metrics of each model on the B1–16 dataset's test sets.

7.4.4 *Classification evaluation results for each model*

According to the definition of the Scottish Environment Protection Agency (SEPA), DO levels below 4 mg/L can damage aquaculture. When hypoxia occurs (DO $<$ 2 mg/L) and persists for a prolonged period, fish, zooplankton, benthic organisms, and other aquatic organisms may suffer fatal harm, leading to deteriorating water quality and loss of ecosystem function (Jasmin *et al.*, 2022). We set three levels based on the DO concentration to reflect the degree of threat to the survival of aquatic organisms, and these threat levels were set as the output of the PKBiLSTM and other models. The details of categorizing DO concentration into three different classes are displayed in Table 7.10.

Table 7.9. The error evaluation metrics of each model on B1–16 dataset's test sets (the bold text in the table indicates the optimal metric achieved on the test set of this dataset among models M1–M10).

		M1	M2	M3	M4	M5	M6	M7	M8	M9	M10
B1	RMSE	0.213	0.216	0.256	0.246	0.266	0.230	**0.207**	0.286	0.235	0.254
	MAE	0.143	0.149	0.183	0.174	0.185	0.163	**0.135**	0.202	0.152	0.151
	NSE	0.931	0.929	0.901	0.909	0.892	0.920	**0.935**	0.876	0.916	0.902
	KGE	0.931	0.925	0.890	0.907	0.870	0.912	**0.937**	0.868	0.898	0.916
	R^2	0.965	0.967	0.960	0.962	0.948	0.965	**0.967**	0.944	0.959	0.950
B2	RMSE	1.275	**1.219**	1.570	1.309	1.901	1.252	1.348	1.460	1.417	1.857
	MAE	0.921	**0.829**	1.078	0.933	1.542	0.812	0.864	0.990	0.911	1.338
	NSE	0.915	**0.923**	0.871	0.911	0.812	0.918	0.905	0.889	0.895	0.820
	KGE	0.935	**0.953**	0.926	0.924	0.781	0.913	0.939	0.941	0.918	0.887
	R^2	0.962	0.962	0.937	**0.963**	0.908	0.959	0.952	0.944	0.947	0.906
B3	RMSE	0.460	0.472	0.583	0.554	0.711	**0.460**	0.561	0.686	0.544	0.601
	MAE	**0.338**	0.346	0.417	0.427	0.532	0.340	0.406	0.513	0.386	0.452
	NSE	0.935	0.932	0.896	0.906	0.845	**0.935**	0.904	0.856	0.909	0.889
	KGE	0.953	**0.959**	0.857	0.950	0.875	0.888	0.881	0.876	0.924	0.855
	R^2	0.968	0.967	0.951	0.956	0.930	**0.970**	0.953	0.930	0.954	0.947
B4	RMSE	**0.485**	0.502	0.530	0.510	0.605	0.486	0.562	0.620	0.557	0.710
	MAE	0.381	0.396	0.413	0.396	0.466	**0.379**	0.436	0.485	0.430	0.564
	NSE	**0.910**	0.904	0.893	0.901	0.861	0.910	0.879	0.853	0.882	0.808
	KGE	0.924	**0.925**	0.908	0.917	0.780	0.896	0.920	0.927	0.900	0.835
	R^2	**0.956**	0.956	0.951	0.951	0.944	0.955	0.941	0.928	0.940	0.921
B5	RMSE	**0.410**	0.458	0.487	0.489	0.584	0.503	0.546	0.650	0.626	0.683
	MAE	**0.318**	0.368	0.384	0.391	0.458	0.389	0.433	0.517	0.511	0.560
	NSE	**0.947**	0.934	0.926	0.925	0.893	0.921	0.906	0.868	0.877	0.854
	KGE	**0.950**	0.837	0.946	0.869	0.866	0.904	0.944	0.822	0.925	0.846
	R^2	0.974	**0.977**	0.968	0.971	0.951	0.971	0.967	0.937	0.966	0.926
B6	RMSE	0.362	0.418	0.465	0.427	0.430	0.428	**0.342**	0.409	0.482	0.456
	MAE	0.276	0.306	0.370	0.316	0.323	0.320	**0.260**	0.310	0.369	0.329
	NSE	0.951	0.935	0.920	0.932	0.932	0.932	**0.957**	0.938	0.914	0.923
	KGE	0.975	0.920	0.909	0.963	0.949	0.953	0.953	0.962	0.947	0.925
	R^2	0.976	0.975	0.971	0.971	0.968	0.973	**0.981**	0.969	0.969	0.966
B7	RMSE	**0.254**	0.261	0.284	0.271	0.524	0.263	0.313	0.305	0.367	0.371
	MAE	0.184	0.188	0.214	0.189	0.380	**0.181**	0.242	0.232	0.271	0.273
	NSE	**0.940**	0.936	0.924	0.931	0.743	0.935	0.908	0.913	0.874	0.871
	KGE	0.956	0.948	0.927	0.931	0.709	0.944	0.945	**0.960**	0.937	0.861
	R^2	**0.972**	0.968	0.965	0.965	0.878	0.969	0.962	0.963	0.938	0.936
B8	RMSE	**0.980**	1.010	1.461	1.472	1.775	1.190	1.167	1.384	1.447	1.366
	MAE	0.738	**0.706**	1.073	1.056	1.339	0.864	0.784	1.013	1.067	0.946
	NSE	**0.905**	0.900	0.790	0.787	0.690	0.860	0.866	0.811	0.794	0.816
	KGE	**0.951**	0.946	0.882	0.890	0.723	0.915	0.885	0.887	0.873	0.883
	R^2	**0.954**	0.952	0.892	0.893	0.838	0.938	0.937	0.903	0.894	0.905
B9	RMSE	**0.290**	0.292	0.320	0.318	0.337	0.332	0.338	0.358	0.374	0.411
	MAE	0.215	**0.196**	0.225	0.213	0.243	0.224	0.241	0.236	0.253	0.265
	NSE	**0.935**	0.934	0.920	0.921	0.912	0.914	0.911	0.900	0.891	0.869
	KGE	**0.961**	0.922	0.919	0.876	0.941	0.877	0.856	0.861	0.888	0.856
	R^2	**0.973**	0.968	0.965	0.963	0.960	0.960	0.961	0.954	0.945	0.935
B10	RMSE	0.228	0.242	0.248	**0.220**	0.307	0.238	0.251	0.326	0.265	0.260
	MAE	**0.152**	0.165	0.175	0.166	0.202	0.174	0.165	0.234	0.191	0.188
	NSE	0.930	0.921	0.917	**0.935**	0.873	0.923	0.915	0.857	0.905	0.909
	KGE	**0.964**	0.924	0.941	0.946	0.846	0.963	0.927	0.823	0.947	0.901
	R^2	0.965	0.961	0.958	**0.974**	0.939	0.967	0.957	0.932	0.954	0.958

(*Continued*)

Table 7.9. (*Continued*)

		M1	M2	M3	M4	M5	M6	M7	M8	M9	M10
B11	RMSE	0.217	**0.209**	0.228	0.241	0.257	0.235	0.226	0.257	0.254	0.272
	MAE	0.143	**0.143**	0.156	0.159	0.177	0.153	0.149	0.171	0.181	0.181
	NSE	0.942	**0.946**	0.936	0.928	0.919	0.932	0.937	0.919	0.921	0.909
	KGE	**0.942**	0.933	0.927	0.912	0.934	0.913	0.942	0.923	0.891	0.885
	R^2	0.972	**0.973**	0.969	0.966	0.960	0.966	0.969	0.959	0.962	0.956
B12	RMSE	**0.305**	0.311	0.462	0.405	0.431	0.336	0.342	0.372	0.420	0.445
	MAE	**0.211**	0.213	0.310	0.285	0.293	0.234	0.255	0.255	0.298	0.357
	NSE	**0.925**	0.922	0.829	0.868	0.851	0.910	0.906	0.889	0.858	0.841
	KGE	0.842	0.865	0.801	0.819	0.806	**0.900**	0.879	0.853	0.809	0.827
	R^2	**0.971**	0.965	0.932	0.939	0.937	0.955	0.955	0.952	0.934	0.939
B13	RMSE	0.334	0.361	0.354	0.364	0.400	**0.313**	0.346	0.326	0.348	0.411
	MAE	0.231	0.265	0.250	0.254	0.286	**0.223**	0.257	0.236	0.236	0.301
	NSE	0.899	0.882	0.886	0.880	0.855	**0.911**	0.892	0.904	0.890	0.847
	KGE	0.849	0.854	0.852	0.857	0.862	**0.890**	0.835	0.876	0.857	0.804
	R^2	0.953	0.944	0.946	0.941	0.931	0.958	0.951	**0.954**	0.947	0.931
B14	RMSE	**0.420**	0.467	0.497	0.525	0.633	0.545	0.478	0.542	0.491	0.530
	MAE	**0.325**	0.372	0.398	0.426	0.511	0.447	0.377	0.448	0.385	0.420
	NSE	**0.835**	0.796	0.768	0.741	0.625	0.721	0.786	0.725	0.774	0.737
	KGE	**0.937**	0.858	0.893	0.880	0.781	0.834	0.881	0.867	0.862	0.875
	R^2	**0.944**	0.903	0.894	0.895	0.848	0.881	0.896	0.872	0.886	0.885
B15	RMSE	**0.750**	0.770	0.848	0.864	1.039	0.909	0.869	1.129	0.988	1.032
	MAE	0.534	**0.523**	0.564	0.616	0.803	0.698	0.565	0.871	0.763	0.815
	NSE	**0.936**	0.932	0.918	0.915	0.877	0.906	0.914	0.854	0.889	0.878
	KGE	0.869	0.875	0.877	0.818	0.864	0.876	**0.887**	0.832	0.854	0.880
	R^2	**0.972**	0.969	0.962	0.966	0.938	0.954	0.958	0.931	0.946	0.938
B16	RMSE	0.278	0.284	0.324	0.308	0.357	**0.277**	0.308	0.296	0.324	0.437
	MAE	0.173	0.192	0.187	0.176	0.259	**0.171**	0.205	0.192	0.194	0.289
	NSE	0.933	0.930	0.909	0.918	0.889	**0.933**	0.917	0.924	0.909	0.834
	KGE	0.909	**0.919**	0.884	0.909	0.842	0.901	0.866	0.919	0.932	0.798
	R^2	0.967	0.967	0.956	0.959	0.949	**0.968**	0.966	0.966	0.954	0.919

Table 7.10. The details of categorizing DO concentration into three different classes.

Threat Degree to the Survival of Aquatic Organisms	DO Concentrations (mg/L)
Light	>4
Mild	2–4
Significant	<2

We set cross-entropy as the loss function. Then, we use metrics such as F1 and recall to comprehensively assess the performance of each model. The weighted average results for each model on test sets B8 and B15 are presented in Table 7.11.

Table 7.11. Weighted average results for each model on test sets B8 and B15.

		Precision	Accuracy	Recall	F1
B8	M1	0.988	0.988	0.988	0.987
	M2	0.987	0.986	0.987	0.985
	M3	0.983	0.967	0.983	0.975
	M4	0.986	0.981	0.986	0.982
	M5	0.950	0.973	0.950	0.960
	M6	0.986	0.984	0.986	0.984
	M7	0.955	0.971	0.955	0.963
	M8	0.963	0.969	0.963	0.966
	M9	0.938	0.969	0.938	0.953
	M10	0.945	0.966	0.945	0.956
B15	M1	0.900	0.896	0.900	0.897
	M2	0.899	0.896	0.899	0.897
	M3	0.897	0.894	0.896	0.893
	M4	0.897	0.893	0.897	0.894
	M5	0.872	0.861	0.872	0.864
	M6	0.890	0.890	0.890	0.890
	M7	0.890	0.886	0.890	0.888
	M8	0.886	0.880	0.886	0.882
	M9	0.889	0.881	0.889	0.884
	M10	0.755	0.569	0.755	0.649

7.5 Discussion

7.5.1 *Classification results analysis*

Figure 7.3 presents the ROC curves and AUC results for M1 on the (a) B8 and (b) B15 test sets.

The ROC curve's horizontal axis represents the false positive rate (FPR), indicating how often negative cases are incorrectly classified as positive. The vertical axis represents the true positive rate (TPR), showing how accurately positive cases are identified. The AUC, or area under the curve, ranges from 0.5 to 1, with a higher value indicating better classifier performance. From Figure 7.3, on the B8 and B15 test sets, the M1 model showed a high AUC of 1 and 0.98 in predicting the severity category of aquatic organisms' survival situations, indicating that the PKBiLSTM model demonstrates good performance in the classification task.

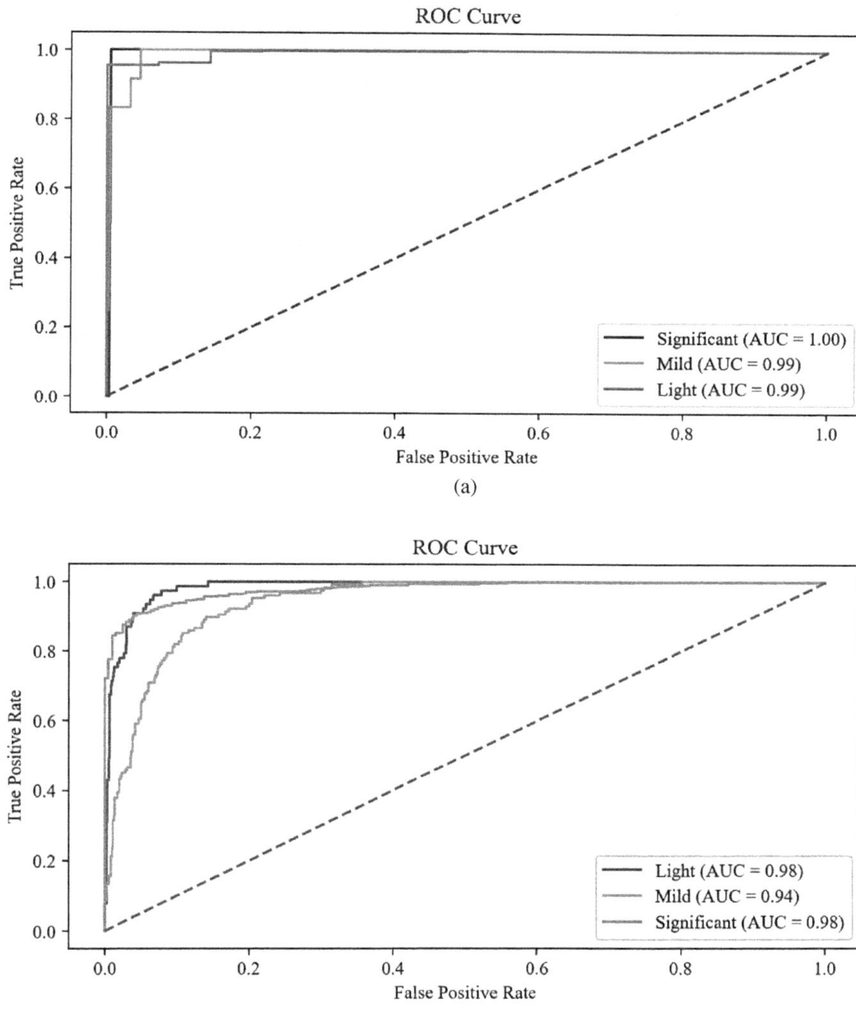

Figure 7.3. The ROC curves and AUC results for M1 on the (a) B8 and (b) B15 test sets.

Figure 7.4 displays stacked histograms of each evaluation metric for each model on the test sets (a) B8 and (b) B15.

From Figure 7.4, the precision of the M1 model on the B8 and B15 test sets is 0.987 and 0.897, respectively, which is 0.51% and 0.33%

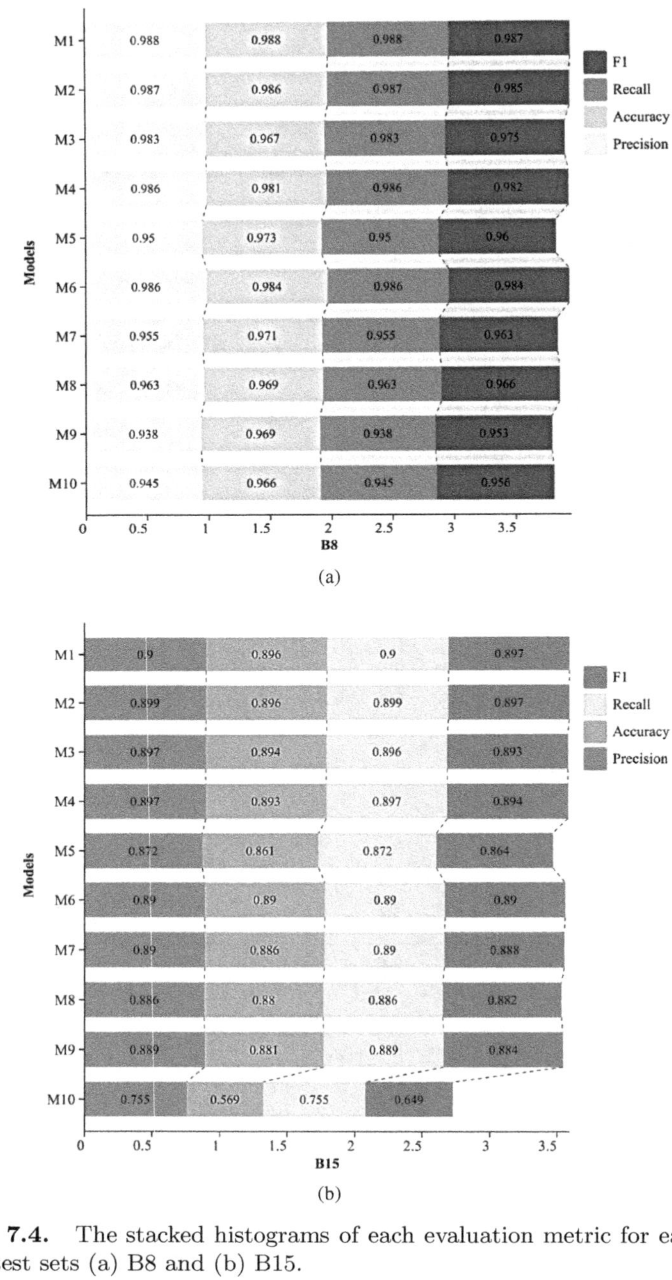

Figure 7.4. The stacked histograms of each evaluation metric for each model on the test sets (a) B8 and (b) B15.

higher than that of the M3 model, indicating that using the PKBiLSTM model to classify DO concentration significantly improves classification accuracy. Meanwhile, on the B8 and B15 test sets, the F1 score of the PKBiLSTM (M1) model is 0.987 and 0.897, respectively, representing an improvement of 3.44% and 1.45% compared to the RF (M9) model. Hence, adopting the PKBiLSTM model to categorize DO concentrations is effective and competitive.

7.5.2 *Regression results analysis*

The RMSE heat map for each model on the test set of each dataset is displayed in Figure 7.5.

From Table 7.9, taking the B4 and B5 test sets as examples, the PKBiLSTM model demonstrates high overall fitting accuracy with NSE of 0.910 and 0.947, respectively. In addition, Table 7.7 shows that the PKBiLSTM model exhibits a higher overall fitting accuracy

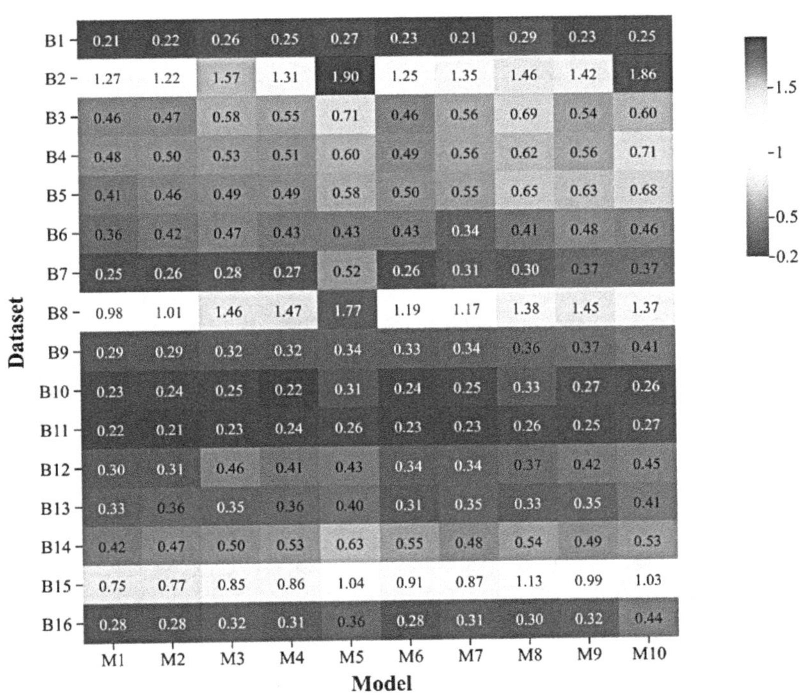

Figure 7.5. The RMSE heat map for each model on the test set of each dataset.

compared to the BiLSTM, LSTM, and GRU models on the test sets of 16 datasets. For instance, on the B16 test set, the RMSE and MAE of the M1 model are 0.278 and 0.173, respectively. Compared to the M4 model, the M1 model shows a decline of 9.74% and 1.7% in RMSE and MAE, respectively. Meanwhile, the KGE of the M1 model on the B9 test set is 0.961, which represents an improvement of 4.37% and 8.84% compared to the B3 and B4 models, respectively.

Moreover, the PKBiLSTM model outperforms other ML models in terms of overall assessment metrics. Figure 7.6 displays the boxplots of (a) RMSE, (b) NSE, and (c) KGE for each model on each dataset's test set.

From Figure 7.6, the average RMSE of the PKBiLSTM model is 0.443, which has declined by 17.20% and 13.98% compared to the M3 and M4 models, respectively. In addition, the average KGE of the PKBiLSTM model is 0.926, which has improved by 2.48% and 2.38% compared to the M6 and M7 models, respectively. The results indicate that compared to other ML models, the PKBiLSTM model

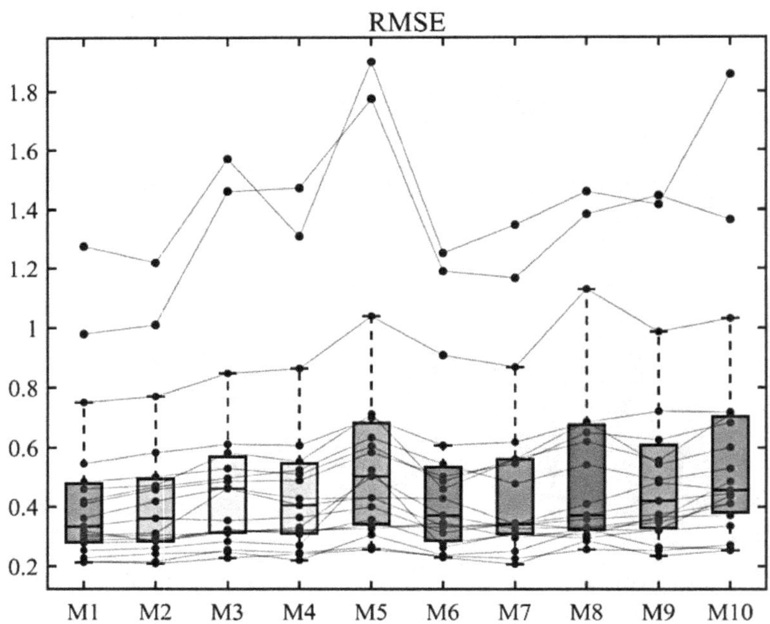

(a)

Figure 7.6. The boxplots of (a) RMSE, (b) NSE, and (c) KGE for each model on each dataset's test set.

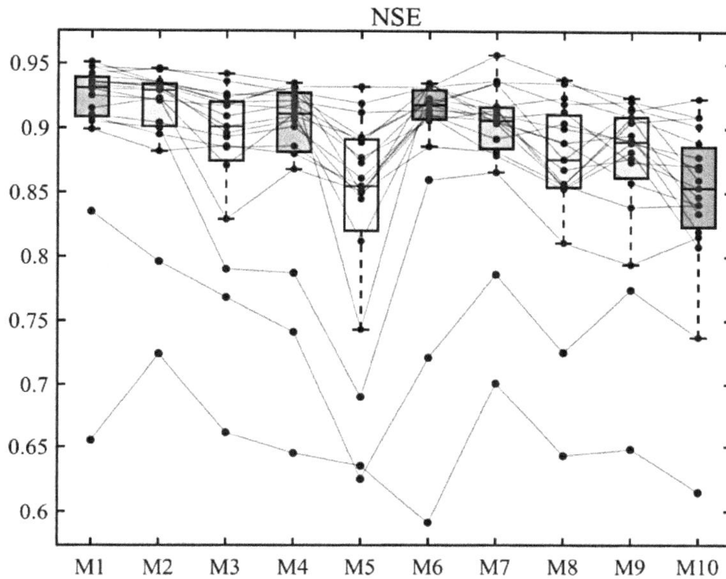

(b)

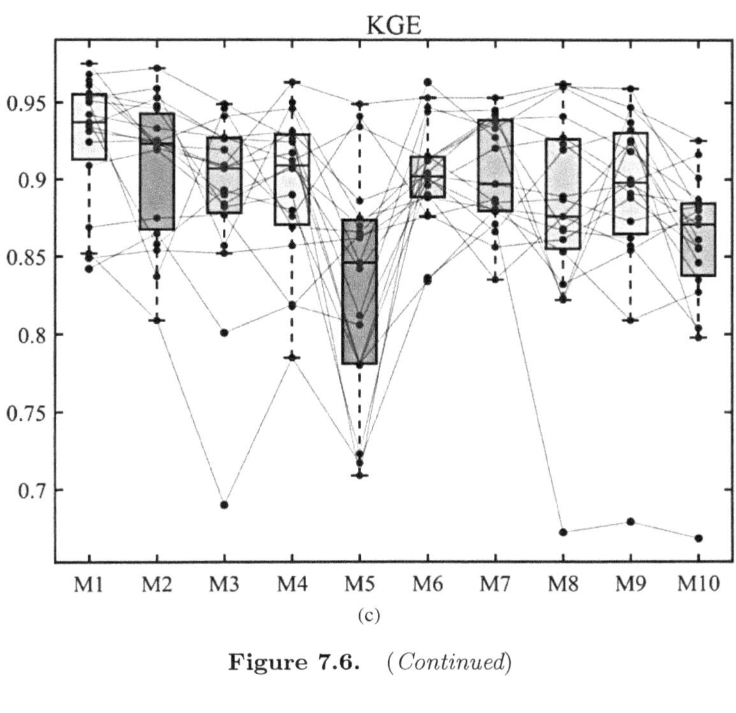

(c)

Figure 7.6. (*Continued*)

possesses better fitting and generalization capabilities when dealing with DO sequences that exhibit strong randomness and nonlinearity.

In addition, the PKBiLSTM model demonstrates superior performance compared to the BiLSTM model. Compared to the BiLSTM model, the PKBiLSTM model shows a closer fit between predicted values and actual values. For example, on the S3 test set, the R^2 of the PKBiLSTM model is 0.968, representing a 0.1% improvement compared to the BiLSTM model. The reason can be attributed to the inclusion of physical and chemical constraints, such as water temperature and DO, in the loss function of the PKBiLSTM model, thus limiting the predicted value. However, the overall high RMSE of PKBiLSTM and other models on the B2 dataset can be attributed to the large amount of noise in the dataset.

Furthermore, the GRU model significantly outperforms the LSTM model in terms of accuracy. Figure 7.7 displays the NSE stacked bar chart for LSTM and GRU models on the B1–B5 test sets.

From Figure 7.7, on the B1 test set, the NSE for the GRU model is 0.91, which has improved by 0.88% compared to the LSTM model. Considering the overall test set B1–B5, the mean NSE values for GRU and LSTM are 0.8974 and 0.9104, respectively. Compared to LSTM, GRU shows a 1.43% improvement in the mean NSE. The reason for this can be attributed to the simpler structure and fewer parameters of the GRU model compared to the LSTM model (Gao *et al.*, 2020).

7.5.3 *Stability testing*

Moreover, considering that the initialization results of deep learning models can have a significant impact on the final accuracy of those models, to further examine the stability of the PKBiLSTM model, we ran the PKBiLSTM and other deep learning models five times each. Figure 7.8 presents the NSE ridgeline chart for five runs of the models M1–M6 on the B1 datasets.

From Figure 7.8, when M1 is run five separate times, the mean value of NSE is 0.9302, an increase of 0.3% and 0.97% compared to the values observed for models M2 and M6. In addition, the NSE standard deviation of the M1 model after five runs is 0.003, which is 40% lower compared to M2. As a result, the M1 model significantly outperforms other deep learning models in terms of stability.

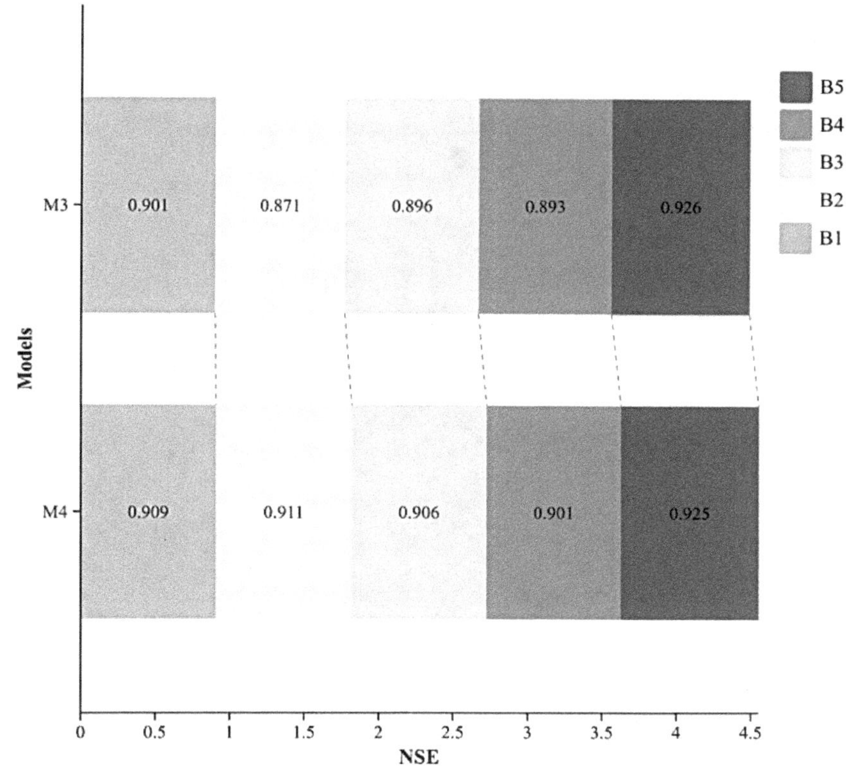

Figure 7.7. The NSE stacked bar chart for LSTM and GRU models on test sets B1–B5.

7.5.4 *Input variable importance analysis*

While PKBiLSTM models outperform other models on multiple datasets, one of the main problems with deep learning models lies in their black-box nature. Deep learning models typically consist of multi-layer neural networks that contain a large number of parameters and complex, non-linear mappings, which makes understanding and interpreting their internal mechanisms challenging. As a result, it is difficult to attribute the observed decrease in DO at a given moment to a specific factor. In this study, SHAP was used to analyze the effects of various internal parameters on DO levels in the water bodies.

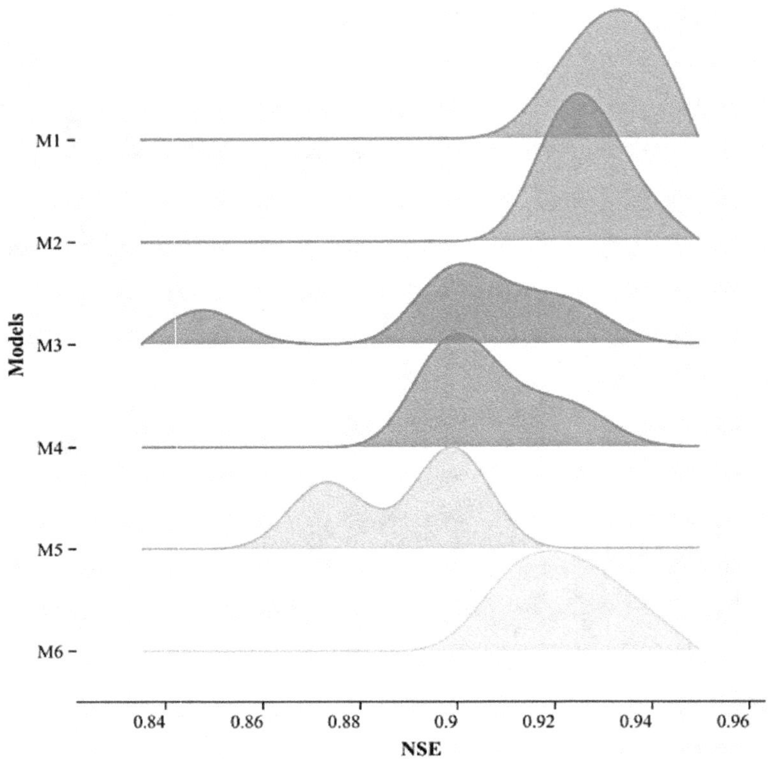

Figure 7.8. The NSE ridgeline chart for five runs of the models M1–M6 on the B1 dataset.

7.5.4.1 *Local interpretability*

Figure 7.9 presents a decision diagram of the DO concentration of the B3 dataset on June 29, 2021, at 00:00.

In the decision graph, the absolute value of the slope of each segment decreases gradually from bottom to top, indicating that while maintaining the same vertical distance, the horizontal distance variation gradually increases, implying a growing impact on the prediction. When the slope is positive, it means that the current value of the corresponding feature contributes positively to the prediction of the instance. Conversely, when the slope is negative, it means that the value of the feature has a negative contribution to the prediction. From Figure 7.9, pH, hour, water temperature, and AN have the

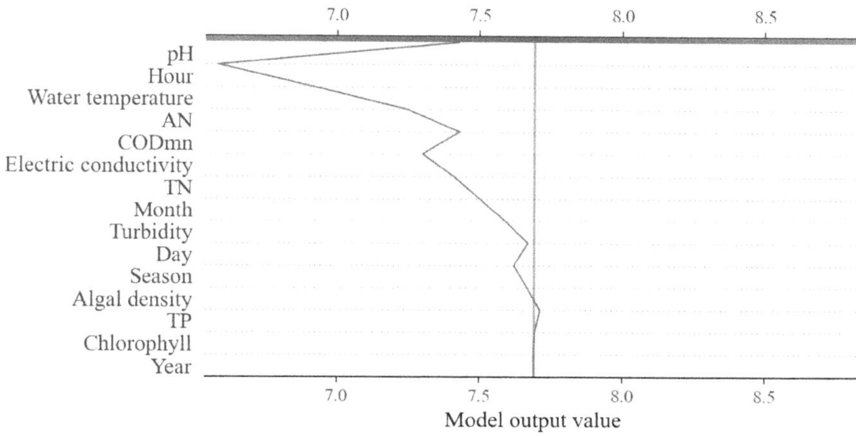

Figure 7.9. The decision diagram of DO concentration of the B3 dataset on June 29, 2021, at 00:00.

highest contributions to the DO concentration of the B3 dataset on June 29, 2021, at 00:00. In addition, pH and AN have a positive contribution to the DO concentration. While chlorophyll, COD_{mn} and algal density have a negative contribution to the DO concentration.

7.5.4.2 *Global interpretability*

Figure 7.10 presents the SHAP summary and bar charts of the input variables for datasets B1–B3.

From Figure 7.10, taking the bar chart in Figure 7.10(a) as an example, it can be observed that the absolute SHAP value of the pH parameter is significantly larger than TP, TN, and other parameters. This indicates that, compared to other internal parameters in the water, pH has the greatest influence on DO. In descending order, water temperature, hour, and season follow the pH parameter in terms of their impact on DO concentrations. In addition, from Figure 7.10(a), it appears that as the pH increases, the SHAP values also increase, indicating that within the pH range of 6.5–10, an increase in pH will lead to an increase in DO. In addition, the concentration of DO decreases with increasing water temperature. In the B2 dataset, similarly, DO concentration increases with an increase in pH. Furthermore, DO decreases with an increase in electric conductivity and turbidity. Lake turbidity reflects the concentration of

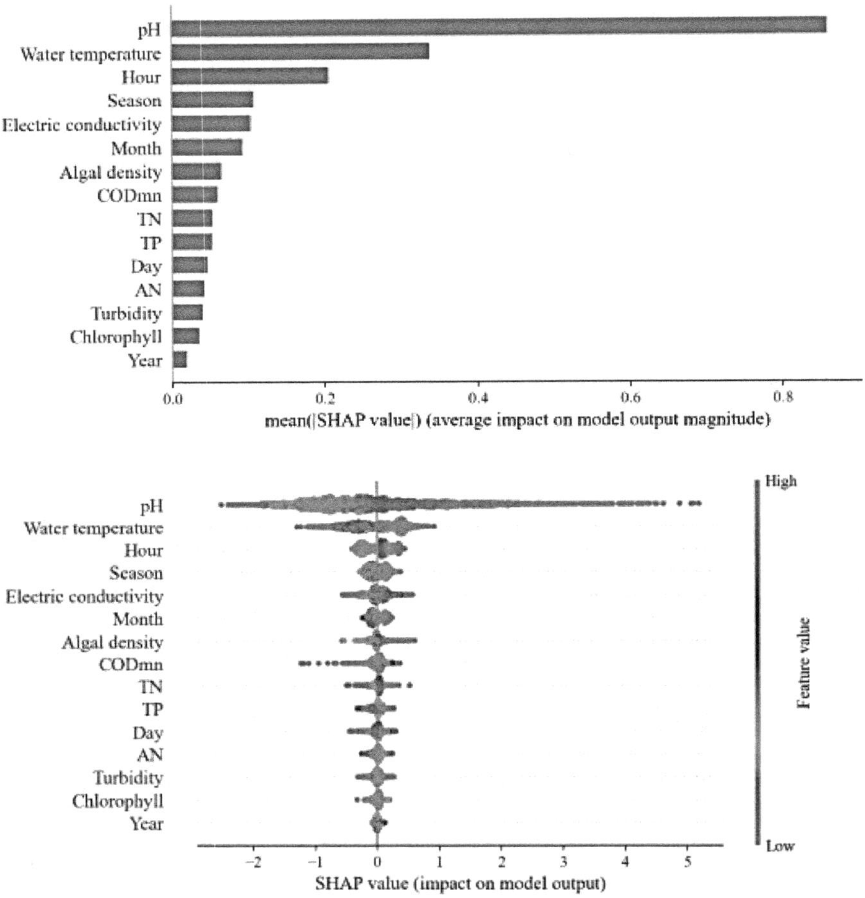

(a)

Figure 7.10. The SHAP summary and bar charts of the input variables in datasets (a) B1, (b) B2 and (c) B3.

suspended particles in the water. As turbidity increases, there is an increase in the absorption and reflection of light by particulate matter in the water, resulting in a decrease in the number of photons entering the water, thereby reducing the efficiency of photosynthesis and suppressing the enhancement of primary productivity in the water body.

The primary reasons for the increase in DO concentration due to elevated pH are listed as follows.

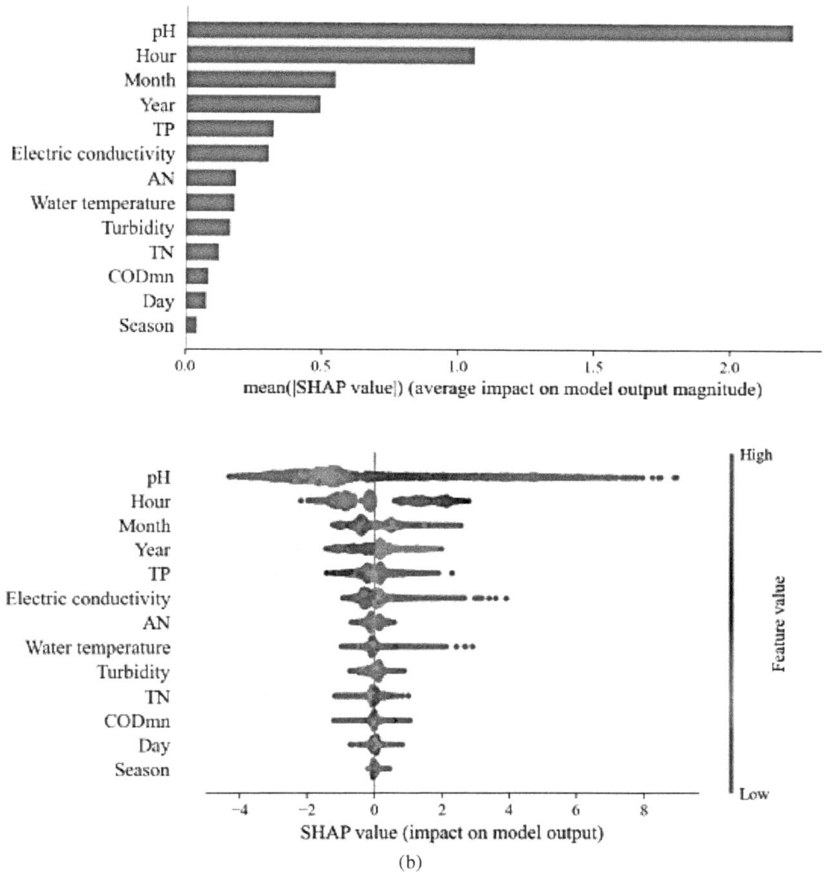

Figure 7.10. (*Continued*)

Firstly, an increase in chlorophyll concentration in natural water bodies indicates a proliferation of phytoplankton, which leads to a higher release of oxygen through photosynthesis. As a result, the DO concentrations in the water increase. The chemical equation is shown in Eq. (7.9):

$$6CO_2 + 6H_2O + E \longrightarrow C_6H_{12}O_6 + 6O_2 \quad (7.9)$$

Furthermore, due to the weak alkalinity of natural water bodies (Lu and Ma, 2020), carbonate exists mainly in the form of HCO_3^-.

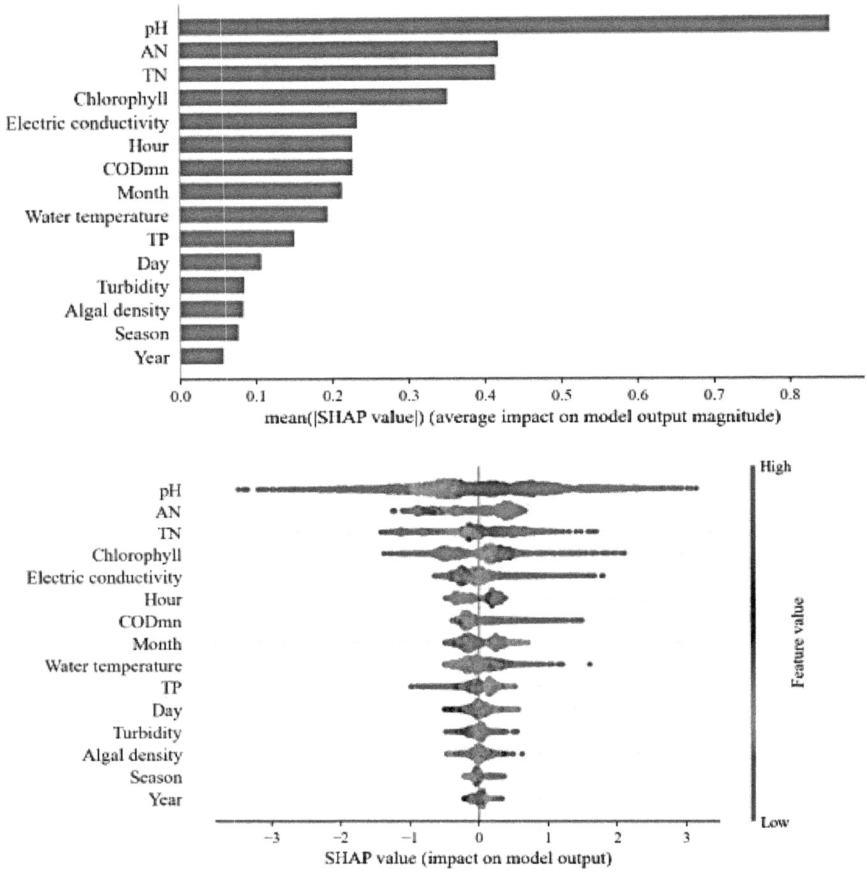

(c)

Figure 7.10. (*Continued*)

The chemical equations are shown in Eq. (7.10)–(7.12):

$$CO_2(g) \rightleftarrows CO_2(\text{aq}) \tag{7.10}$$

$$CO_2(\text{aq}) + H_2O \rightleftarrows H_2CO_3 \tag{7.11}$$

$$H_2CO_3 \rightleftarrows HCO_3^- + H^+ \tag{7.12}$$

HCO_3^- is unstable and can ionize into CO_2, CO_3^{2-}, and H_2O. The chemical equation is shown in Eq. (7.13):

$$2HCO_3^- \rightleftarrows CO_2 + CO_3^{2-} + H_2O \tag{7.13}$$

As a result of photosynthesis, the reduction of CO_2 causes the ionization equation to shift to the right, leading to a decrease in HCO_3^- concentration. The generated CO_3^{2-} then undergoes hydrolysis, producing HCO_3^- and OH^-. The chemical equation is shown in Eq. (7.14):

$$CO_3^{2-} + H_2O \rightleftarrows HCO_3^- + OH^- \tag{7.14}$$

Therefore, the increase in pH leads to an increase in DO concentration.

In summary, within the Taihu Lake basin, in the pH range of 4–11, the DO concentration increases with an increase in pH. In addition, with an increase in conductivity and TP, the concentration of DO decreases.

Subsequently, to further determine the optimal input variables for DO concentration prediction, this study employed RMSE and NSE to analyze and compare the DO prediction performance of different combinations of input variables. Table 7.12 displays the results and abbreviations for each combination of input variables.

Table 7.13 displays the evaluation metrics for each input combination for five separate runs on the B1 test set.

Figure 7.11 presents the (a) NSE and (b) RMSE boxplots of the B1 test set, run five times each, for different combinations of inputs.

From Figure 7.11, it appears that when the input variable combination is K1 (water temperature and pH), the average NSE and RMSE on the B1 test set are 0.9230 and 0.2254, respectively. However, when the input variable combination is K3 (water temperature, pH, conductivity, and TN), the average NSE and the average RMSE are 0.9320 and 0.2120, respectively. Compared to K1, the input combination K3 shows an increase of 0.97% in the average NSE and a

Table 7.12. The results and abbreviations for each combination of input variables.

Number	Input variable	Number	Input variable
K1	A1+A2	K5	A1+A2+A3+A6+A7
K2	A1+A2+A6	K6	A1+A2+A8
K3	A1+A2+A6+A7	K7	A1+A2+A9
K4	A1+A2+A3	K8	A1+A2+A10

Table 7.13. The evaluation metrics for each input combination for five separate runs on the B1 test set.

Number		RMSE	NSE
K1	Mean	0.2254	0.9230
	Std	0.0059	0.0043
K2	Mean	0.2136	0.9310
	Std	0.0032	0.0021
K3	Mean	0.2120	0.9320
	Std	0.0043	0.0027
K4	Mean	0.2152	0.9302
	Std	0.0045	0.0029
K5	Mean	0.2136	0.9306
	Std	0.0067	0.0043
K6	Mean	0.2166	0.9290
	Std	0.0061	0.0039
K7	Mean	0.2198	0.9266
	Std	0.0055	0.0035
K8	Mean	0.2182	0.9276
	Std	0.0079	0.0050

reduction of 5.94% in the average RMSE. Furthermore, the impact of different input variable combinations on model accuracy aligns with the results from the SHAP analysis. For instance, compared to K2, incorporating total nitrogen (TN) in K3 led to a 0.11% increase in the average NSE value and a 0.75% decrease in the average RMSE value. This indicates a positive effect of TN in enhancing the model's accuracy, which is consistent with the findings of the SHAP analysis. Furthermore, it is not necessarily the case that the accuracy of the model increases with an increase in input variables. Taking K3 and K5 as examples, when the combination of input variables is K5, the model's NSE mean value is 0.9306, which is a decrease of 0.15% compared to K3. This indicates that with an increase in input variables, there is a significant decrease in the fitting accuracy of the model. In summary, when the input variables are water temperature, pH, TN, and conductivity, the B1 dataset achieved the highest accuracy.

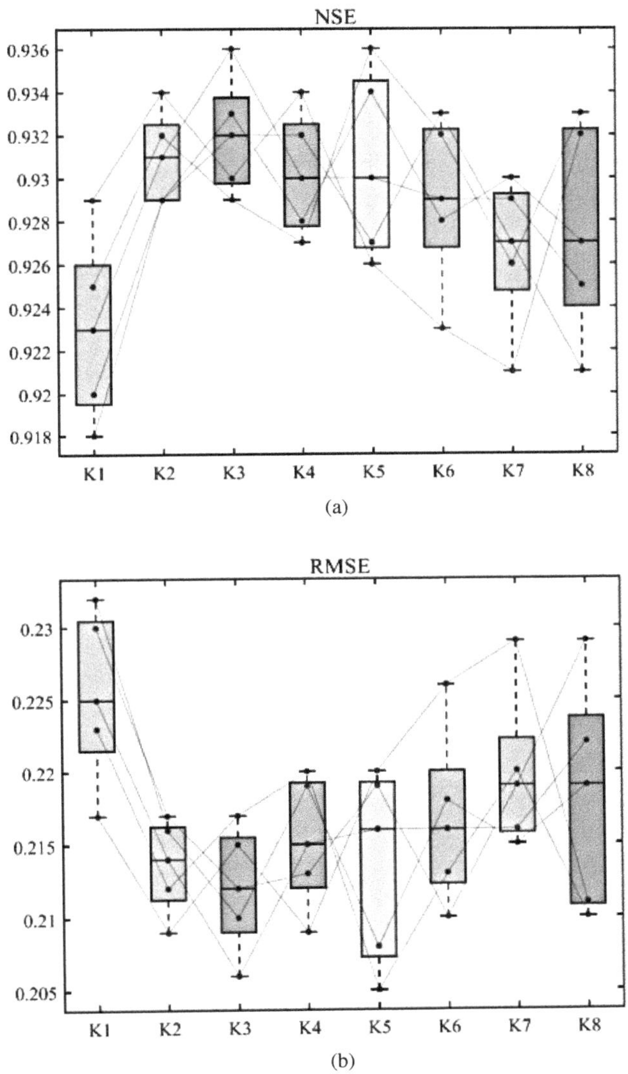

Figure 7.11. The (a) NSE and (b) RMSE boxplots of the B1 test set run.

7.5.5 Model implications

In recent years, as human activities continue to grow, the discharge of pollutants into rivers, lakes, and oceans has increased exponentially. This exponential rise in pollution has resulted in the eutrophication of water bodies, causing detrimental effects on their ecological balance. In addition, the frequency of natural disasters is increasing, especially harmful algal blooms (HABs) and red tides (Tewar *et al.*, 2022; Stump *et al.*, 2016; Yuan *et al.*, 2023; Ai *et al.*, 2023). Accurate water quality prediction enables policymakers, water authorities, and environmental agencies to make informed decisions regarding water allocation, pollution control strategies, and sustainable water use. Nowadays, scholars commonly utilize process-driven and data-driven approaches to predict DO concentrations. While these approaches have their advantages in DO modeling and prediction, they also face drawbacks and challenges. For instance, process-driven methods, which focus on simulating DO levels using complex mathematical equations and models, provide good interpretability and aid in understanding the impacts of various factors on water quality. However, they may encounter difficulties in data collection and processing, as well as the complexity and time-consuming nature of the modeling process. On the other hand, data-driven methods emphasize the collection and analysis of extensive observational data, employing statistical and ML techniques to establish predictive models. Nonetheless, data-driven methods often lack a deep understanding of the underlying mechanisms behind water quality changes, struggle to comprehend the intricate physical and chemical variations within the water body, and exhibit limited interpretability of model results, thereby placing restrictions on their further application.

Hence, in this study, we considered the physical and chemical changes of DO within water bodies and incorporated them as prior knowledge in the loss function of a deep learning model to constrain the predicted values. In addition, considering the ability of the BiLSTM model to effectively utilize past and future information for training and its strong nonlinear fitting capability, we modified the loss function of the BiLSTM model and proposed the PKBiLSTM model. Subsequently, considering the poor water quality in the Taihu Lake Basin, we focused our study on 16 hydrological monitoring sites within the basin. To evaluate the effectiveness of the

PKBiLSTM model, we compared its performance with nine other classic models, including CatBoost. Then, to address the need for interpretability in deep learning models, we introduced the SHAP method to assess the contributions of various parameters, such as water temperature, TN, and TP, to DO in water bodies. By using SHAP, we can enhance the interpretability of the model and provide a clearer understanding of the underlying relationships. The results indicate that the PKBiLSTM model outperforms other models such as CatBoost in overall performance for the Taihu Lake Basin, demonstrating excellent robustness and generalization capabilities. In addition, compared to factors such as water temperature, conductivity, and turbidity, pH has the greatest impact on DO. Within the pH range of 6.5–10, as the pH increases, the concentration of DO also increases.

7.5.6 *Model deficiencies*

Despite the impressive performance of the PKBiLSTM model across different datasets, showcasing its robustness and generalizability, as well as its ability to elucidate the impact of individual input variables on DO, this research still has a few areas that require improvement:

(1) In this study, we only considered and examined the effects of various parameters within the water body on the variation of DO. Nonetheless, the variation of DO is influenced by multiple factors such as human activities and geological environments, and it has a significant degree of uncertainty. In subsequent studies, we will further refine this model to account for these additional factors and improve its reliability.
(2) In this study, the Taihu Lake Basin served as the sole research object. However, future research will gradually apply the PKBiLSTM model to other basins to further evaluate its generalizability and robustness.

7.6 Conclusions

In recent years, the rapid acceleration of industrialization has resulted in severe degradation of river water environments. Establishing reliable water quality models is of paramount importance for predicting changes in river water quality and implementing effective decision-making. In this study, we propose the PKBiLSTM model to simulate DO in the entire Taihu Lake Basin. Additionally, we incorporate SHAP values to analyze the impact of various internal parameters within each water body on DO. The findings demonstrate that the PKBiLSTM model outperforms other existing models across 10 datasets, showcasing its robustness and remarkable generalization capabilities. Furthermore, our analysis reveals that pH, among numerous internal parameters within the water bodies, exerts the most significant influence on DO. Notably, within the pH range of 6.5–10, an increase in pH corresponds to a gradual rise in DO concentration. Our findings contribute to a deeper understanding of water quality dynamics and provide valuable insights for environmental management and conservation efforts in the studied region.

However, there are still a few shortcomings in this study. Firstly, DO is closely related to factors such as human activities, but this study only reveals the effect of various internal factors within the water body on DO. Secondly, the study is limited to the entire Taihu Lake Basin. In future research, we will further improve the model and explore the impact of human activities and other factors on DO, hoping that future scholars can further refine it.

Chapter 8

Assessment and Strategies for Water Supply Security Risks in the Estuarine City of Shanghai under Normalized Extreme Climate Conditions

Facing the escalation of extreme climate events, estuarine delta cities like Shanghai grapple with significant water supply challenges. This study employs system dynamics and dynamic adaptive policy pathways to assess Shanghai's water supply security risks and responsive strategies from 2022 to 2050, amid extreme climate conditions. Utilizing data from 2000 to 2021, we constructed a system dynamics model to predict future water demand under various development modes. Focusing on the unusual 2022 drought in the Yangtze River Basin, we simulated 15 scenarios, including economy, population, water efficiency, and reservoir levels, to identify the extent and timing of potential water supply risks and then proposed pertinent dynamic adaptive strategies to address them. Our findings suggest that the 2022 drought significantly reduced Shanghai's water supply capacity, leading to a notable deficit. Under scenarios of accelerated economic growth, water supply security risks are heightened, with projections indicating a reduction in days of supply to merely 33–67 days and escalating water shortages of 592–896 million m^3 by 2050. Short-to-medium-term recommendations include optimizing both local and transit water resources, strengthening emergency water reserves, enhancing water use efficiency, and maintaining stable reservoir water levels. For the long term, expanding water storage infrastructure and promoting integrated water resource management within the Yangtze River Delta is key to establishing a resilient and diversified water supply system, effectively mitigating future water security risks. This study provides a scientific basis and reference for the sustainable management of water resources in estuarine cities confronting normalized extreme climate conditions. It offers valuable insights for policymakers and actionable suggestions for urban planners.

8.1 Introduction

In recent years, the global rise in both the frequency and intensity of extreme climate events (normalized extreme climate) has significantly disrupted water distribution and supply patterns (IPCC, 2022; Yao *et al.*, 2023; Zscheischler *et al.*, 2018), posing serious threats to the sustainable development of regional water resources (Zhang *et al.*, 2023). According to the United Nations, the global urban population is projected to increase by 2.3 billion over the next 30 years, leading to an 80% surge in urban water demand (Flörke *et al.*, 2018). This rapid growth, along with intensifying climate conditions, indicates a substantial rise in urban water shortages (Omer *et al.*, 2020; Pokhrel *et al.*, 2021). By 2050, an estimated 1.5–2.5 billion people may face water scarcity each year, potentially doubling current figures (Rahman *et al.*, 2023). Therefore, addressing the interconnected challenges of climate change, population growth, and economic development is crucial for securing regional water sustainability (Ahmadi *et al.*, 2020; Cheng and Chen, 2017).

Water security, a critical concern in 21st-century global research, refers to maintaining an acceptable level of water-related risk while ensuring sufficient water quantity and quality for human and ecosystem well-being, supporting livelihoods, national security, and health (Bakker, 2012; Li *et al.*, 2023). Water supply security (WSS), often used interchangeably with concepts of water scarcity, pertains to discussions about water sufficiency and challenges related to water supply (Biswas and Rahman, 2023). Risk assessments in this field typically involve establishing an indicator system within a framework. Common methods include fuzzy evaluation (Liu *et al.*, 2017), probabilistic statistics (Jabari *et al.*, 2020; Tang *et al.*, 2023), spatial overlay (Dou *et al.*, 2021), system dynamics (SD) (Sušnik *et al.*, 2012; Zhou *et al.*, 2023), and neural networks (Mo *et al.*, 2024; Xu *et al.*, 2022). A typical assessment framework involves analyzing the balance between water supply and demand (Liu *et al.*, 2021; Wang *et al.*, 2019). After determining water supply and demand, risk performance indexes are used for further assessment, such as the water stress index (WSI) (Hanasaki *et al.*, 2013; Huang *et al.*, 2021), the vulnerability index (Yang *et al.*, 2022), the poverty index (Sullivan *et al.*, 2003), and the drought risk index (Nasrollahi *et al.*, 2018).

Dynamic adaptive frameworks such as dynamic adaptive policy pathways (DAPP), policymaking, and scenario planning are developed to navigate climate change uncertainties that vary over time (Haasnoot *et al.*, 2013; Tian *et al.*, 2022). The DAPP approach, which integrates adaptation pathways and adaptation policymaking approaches, facilitates more effective adaptation planning, implementation, and response through forward planning and alignment with site-specific development priorities (IPCC, 2022). In this method, the plan is designed to sort a series of possible actions (paths) over time, where each possible adaptation action has an expiry date, called a tipping point. This tipping point represents the moment when climate change causes the current management strategy to fail to achieve its goal, necessitating a shift to alternative approaches. This highlights the importance of management based on risk sequences and from the perspective of decision-makers (Haasnoot *et al.*, 2013; Kwadijk *et al.*, 2010; McKay *et al.*, 2022). The DAPP method supports the dynamic exploration of different associated uncertainties, the linking of short-term goals with long-term goals, and the determination of short-term actions while preserving future options. It has been widely used in water resources planning, coastal risk management, and other fields (Babaeian *et al.*, 2023; Hamlington *et al.*, 2023; Toimil *et al.*, 2021). Therefore, the DAPP method is a good choice for dealing with the impacts of extreme climate events and uncertainties regarding WSS. Even if the impacts of extreme climate change in the future are not as expected, it is still beneficial to prepare for the worst-case scenario (Tian *et al.*, 2023).

DAPP must be combined with quantitative computational methods to determine the magnitude and timing of risk occurrences and to take timely countermeasures. The SD method aligns well with the needs of DAPP. It can integrate hydrological, social, economic, and environmental factors and management policies in WSS systems, providing a deep understanding of the behavior of complex systems and their responses to strategies (Liu *et al.*, 2022a; Sušnik *et al.*, 2012; Zhou *et al.*, 2023). In risk management, decision-makers often need to pay attention to the future dynamics of policies and their potential long-term risks (Kotir *et al.*, 2016). Compared to other single-equation econometric models, which are more suitable for short-term quantitative studies and predictions, SD is more suitable for qualitative and quantitative analysis of systems, and more accurate medium- and long-term predictions can be made through scenario analysis and dynamic simulation (Sun *et al.*, 2017). Although the SD model can represent the causal relationship between variables, it usually requires other tools to describe the simulation results for reference. Combining the SD with the DAPP method can not only simulate the effects of different policies but also provide an intuitive implementation path for policymaking, help decision-makers grasp the long-term effects and potential risks, and promote flexible and effective decision-making in a changing environment.

The Yangtze River Delta, one of China's most economically vibrant areas, contributes nearly a quarter of the national gross domestic product (GDP) on less than 4% of the land. Shanghai, the region's leading city at the mouth of the Yangtze River, relies on the river for 80% of its freshwater. Recent increases in extreme weather events and rising sea levels have intensified salinity intrusion at the Yangtze Estuary, undermining the reliability of crucial water reservoirs (Cheng *et al.*, 2018; Zhu *et al.*, 2020). The unprecedented drought of 2022, the worst since records began in 1961 (Xia *et al.*, 2022), compounded by the impacts of typhoons, has further strained Shanghai's water supply, posing significant risks to the city's WSS. Addressing WSS risks and developing dynamic adaptive strategies are crucial for Shanghai's evolution into a modern socialist international metropolis and for enhancing the overall strength and international competitiveness of the entire Yangtze River Delta.

In response to the water security challenges facing delta regions, this study centers on Shanghai and adopts a new framework combining SD and DAPP methods to assess WSS risks from 2022 to 2050. The purpose is to determine the extent and timing of potential water deficits, providing essential data for effective water management. Objectives include: (1) analyzing Shanghai's water usage from 2000 to 2021 and predicting future demand under various development modes; (2) assessing supply capacity under extreme drought conditions; (3) computing WSI, days of supply available (DSA), and water shortage amount (WSA) under different scenarios; and (4) proposing dynamic, adaptive strategies based on simulation results. Although Shanghai serves as the primary focus, the proposed framework for WSS risks offers a reference for sustainable water management in other delta cities worldwide, particularly those facing the impacts of climate change.

8.2 Materials and Methods

8.2.1 *Study area*

Shanghai, China's largest city and the third largest globally, spans 6,340.5 km^2 and comprises 16 districts (Figure 8.1(a)). Located at the mouth of the Yangtze River, the city is bordered by the river to the north, the East China Sea to the east, Hangzhou Bay to the south, and the provinces of Jiangsu and Zhejiang to the west. Shanghai has a subtropical monsoon climate, with an average rainfall of 1,044.1 mm and evaporation of 1,136.7 mm in 2022, according to the Shanghai Meteorological Bureau. The city's water resources comprise local and transit sources (Table 8.1), with the multi-year average local water resource accounting for 0.27% of the total and transit sources constituting 99.73%. The water supply has generally increased over the past decade.

Shanghai is a typical water-scarce city due to water quality issues and is considered one of the cities most at risk of severe water crises internationally (Kelly, 2022). Historically, over 80% of the city's water was sourced from Lake Tai via the Huangpu River (Finlayson *et al.*, 2013). However, increasing pollution and eutrophication have complicated the maintenance of a clean water supply from Lake Tai. In response, Shanghai has developed several reservoirs at the Yangtze River mouth to access its high-quality freshwater, making the Yangtze Estuary and Huangpu River the city's two primary water sources, contributing 80% and 20% to the water supply, respectively (Figures 8.1(b) and 8.1(c)). Since the completion of the Jinze Reservoir in December 2016, Shanghai has adopted a raw water

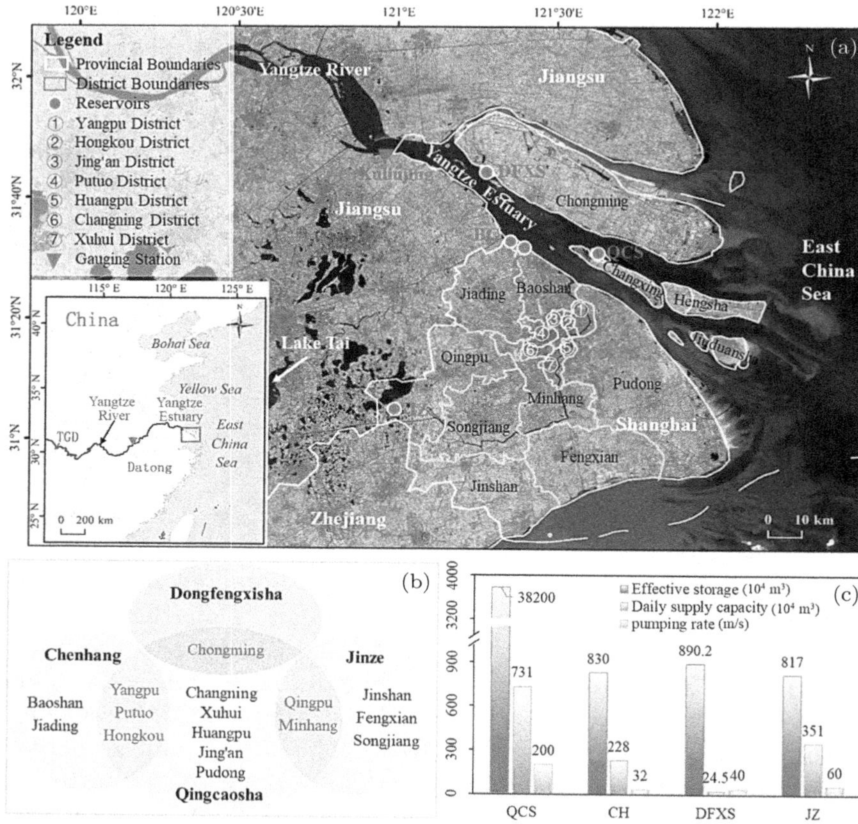

Figure 8.1. Location of Shanghai (a) and supply areas (b) and characteristics (c) of major water supply reservoirs: Qingcaosha (QCS), Chenhang (CH), Dongfengxisha (DFXS), and Jinze (JZ). TGD means the Three Gorges Dam.

supply pattern of "dual rivers and four reservoirs supply." The three reservoirs at the mouth of the Yangtze River (Qingcaosha, Chenhang, and Dongfengxisha) primarily serve as a water supply for central Shanghai and the northeastern areas. In contrast, the Jinze Reservoir on the upstream Huangpu River mainly supplies raw water to the five southwestern districts of Shanghai (Figures 8.1(a) and 8.1(b)). Although the Qingcaosha Reservoir shares a small overlap in its water supply area with the Chenhang and Jinze reservoirs, there is no interconnection between the four reservoirs. However, the Shanghai

Table 8.1. Statistics of water resources in Shanghai.

		Multi-year average		2010–2021 average	
Water type	Item	Quantity (10^8 m^3)	Proportion (%)	Quantity (10^8 m^3)	Proportion (%)
Local resources	Surface runoff	24.15	0.26	36.79	0.38
	Available groundwater	1.42	0.01	0.07	0.00
	Total	25.57	0.27	36.86	0.38
Transit resources	Taihu basin	106.6	1.13	176.32	1.81
	Yangtze river	9335	98.6	9543.75	97.81
	Total	9441.6	99.73	9720.07	99.62

Note: The amount of groundwater that can be extracted after 2015 is the actual amount of groundwater used in that year.

Municipal Water Supply Planning (2019–2035) has proposed to interconnect the raw water systems of Qingcaosha and Chenhang reservoirs in the Yangtze Estuary and the upstream Huangpu River to achieve mutual support and complementarity.

By the end of 2021, Shanghai's permanent population reached nearly 24.9 million, with a GDP of approximately 4.32 trillion RMB and an annual water withdrawal of 10.58 billion m^3, according to the Shanghai Municipal Statistics Bureau. The city's water extraction capacity of 13.345 million m^3/d is typically sufficient to meet demand. Although saltwater intrusion into the Yangtze Estuary occurs annually, it is usually brief and mild, posing a minimal threat to the water supply system. Under normal conditions, the Chenhang, Qingcaosha, and Dongfengxisha reservoirs can independently sustain water supply for up to 5, 68, and 26 days, respectively, without the need for additional extraction. However, the exceptional drought of 2022 severely intensified saltwater intrusion, leading to longer periods without extraction and significantly challenging Shanghai's water supply system.

8.2.2 *Data sources*

This study utilizes socio-economic and hydrological data for Shanghai and the Yangtze Estuary. Socioeconomic data, including population figures, added values for industrial and tertiary industries, and irrigation areas, were derived from the *Shanghai Statistical Yearbook* from 2000 to 2022. Water use data, spanning agricultural, industrial, domestic, urban public, and eco-environmental categories, were sourced from the *Shanghai Water Resources Bulletin* and the National Bureau of Statistics over the same period.

Hydrological information, covering monthly inflows, outflows, and water levels at the Three Gorges Reservoir (TGR), as well as runoff at Datong and Xuliujing gauging stations (Figure 8.1(a)), was obtained from the *Yangtze River Bulletin* for the years 2018 to 2022. Datong station, positioned 600 km upstream of the Yangtze Estuary, is the final mainstream control station in the lower section of the Yangtze River. The daily discharge data for Datong was provided by the Yangtze River Maritime Administration. Information about key reservoirs providing water to the megacity was mainly sourced from regional water management departments.

8.2.3 *Methodology*

As extreme climate events become more common, water shortages emerge as a critical threat to Shanghai's WSS. One way to assess shortage risks is through supply–demand balance analysis (Li *et al.*, 2022). Our methodology synergizes SD with the DAPP approach, not to predict future extreme climate events, but to assess the system's response to past typical events, such as the 2022 drought. This approach simulates the supply–demand balance, assessing risks via the WSI, DSA, and WSA. Additionally, it proposes dynamic adaptive strategies, encompassing engineering and non-engineering measures, to mitigate risks. Water demand depends on the scale of socio-economic development and water use efficiency, both of which are influenced by policy changes. In Shanghai, the primary supply risk is saltwater intrusion, exacerbated by reduced runoff, sea-level rise, and storm surges, leading to decreased freshwater availability from estuarine reservoirs (Figure 8.2(a)).

8.2.3.1 *Water demand prediction*

(1) **SD model**

To predict Shanghai's water demand from 2022 to 2050, an SD model using Vensim software was constructed. This approach, developed by

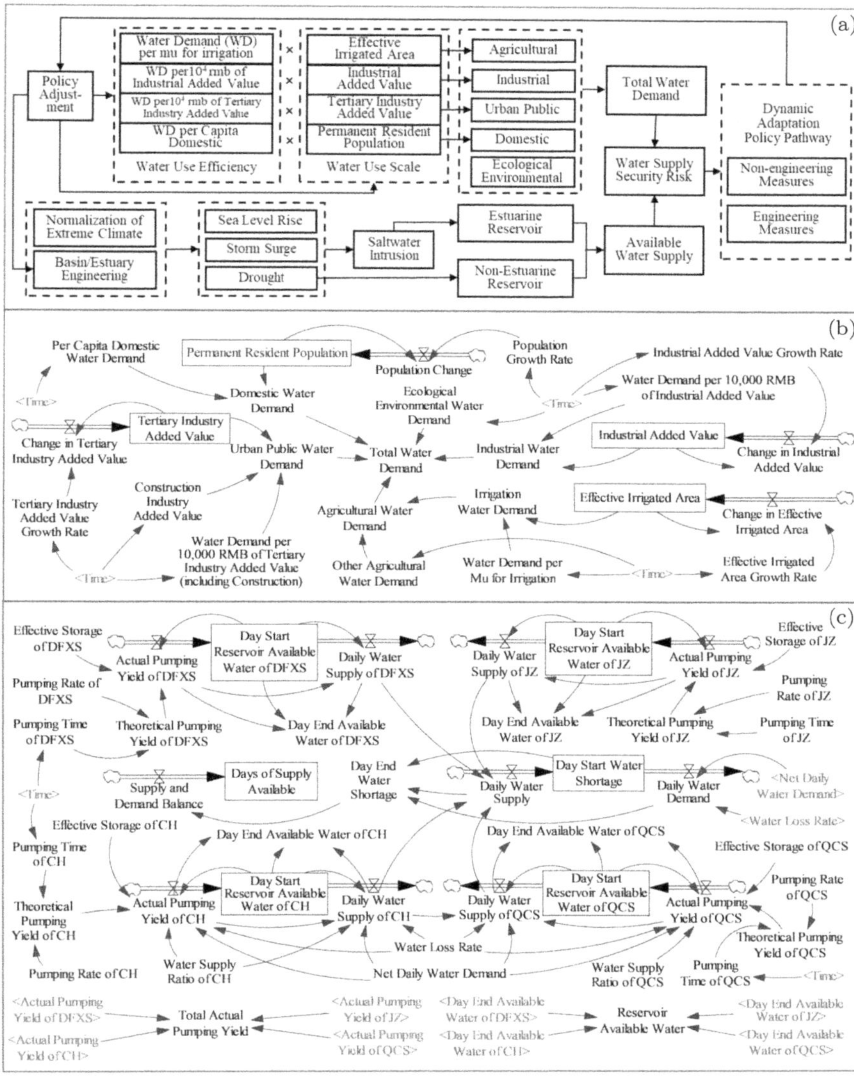

Figure 8.2. Water supply security risk research methods: (a) water supply security risk research framework; (b) stock-flow graph of the system dynamics model for predicting water demand in Shanghai; (c) stock-flow graph for freshwater resources supply–demand balance analysis of Shanghai's reservoirs during saltwater intrusion periods. DFXS, JZ, CH, and QCS represent Dongfengxisha, Jinze, Chenhang, and Qingcaosha Reservoir, respectively.

Jay Forrester at MIT in the 1950s, integrates qualitative and quantitative analyses to address complex, nonlinear problems through simulation, emphasizing interconnectedness and dynamic change. The model considers socio-economic factors impacting water demand across Shanghai's 16 districts, covering sectors such as agriculture, industry, domestic, public urban, and eco-environment. Its variables and structure are visualized in Figure 8.2(b).

Model variable formulas were derived through regression analysis, using historical data from 2000 to 2021. Due to the COVID-19 pandemic outbreak after 2019, some data, such as industrial and tertiary industry added value and water use, showed significant fluctuations and were therefore excluded from the fitting formula.

(2) **Model validation**

To validate the model's accuracy, tests were conducted on its predictive performance across domestic, urban public, agricultural, industrial, and eco-environmental water demand. The relative error was calculated using Eq. (8.1), with an error below 10% deemed acceptable (Sun *et al.*, 2017).

Although the relative error for the urban public water demand in 2020 was high due to the COVID-19 pandemic, average relative errors for the selected test variables ranged from 0.99% to 7.66%. The overall average error is 3.95%, indicating the model's reliability (Table 8.2 and Figure 8.3).

$$\mathrm{RE} = |(M_i - R_i)/R_i| \times 100\% \tag{8.1}$$

where RE is the relative error, and M_i and R_i are the model and real values, respectively.

(3) **Sensitivity analysis**

Sensitivity analysis quantitatively evaluates how changes in input variables affect a primary indicator, allowing for ranking these variables based on their relative importance in achieving evaluation objectives (Zhang *et al.*, 2021). In this study, indicator sensitivity is defined as the degree to which water demand reacts to unit changes in system control indicators. To minimize the impact of randomness, the average variation in water demand for the years 2035 and 2050 under scenarios with ±5% and ±10% indicator fluctuations depicts

Table 8.2. Real water demand values, model simulated values, and relative errors (RE) (unit: 10^8 m^3).

Time	Total water demand			Domestic water demand			Urban public water demand		
	Real value	Model value	RE (%)	Real value	Model value	RE (%)	Real value	Model value	RE (%)
2012	116.00	116.63	0.54	13.27	13.46	1.44	12.33	12.10	1.85
2013	123.20	114.56	7.01	13.71	13.67	0.32	12.03	12.23	1.70
2014	105.90	111.36	5.16	12.75	12.69	0.50	11.61	12.31	6.06
2015	103.80	109.46	5.46	12.51	12.89	3.01	11.63	12.34	6.11
2016	104.80	107.66	2.73	13.15	13.09	0.48	11.97	12.32	2.88
2017	104.80	105.94	1.09	13.23	13.29	0.44	11.34	12.24	7.95
2018	103.40	104.28	0.85	13.26	13.49	1.73	11.23	12.12	7.94
2019	100.90	102.78	1.87	13.50	13.69	1.43	10.69	11.96	11.87
2020	97.50	101.24	3.83	13.89	13.90	0.04	9.66	11.76	21.72
2021	105.80	99.74	5.73	14.03	14.10	0.50	10.62	11.52	8.50

Time	Agricultural water demand			Industrial water demand			Eco-environmental water demand		
	Real value	Model value	RE (%)	Real value	Model value	RE (%)	Real value	Model value	RE (%)
2012	17.50	16.76	4.23	72.90	73.61	0.97	0.70	0.70	0.00
2013	16.30	16.46	0.97	80.40	71.40	11.19	0.80	0.80	0.00
2014	14.60	16.21	11.01	66.20	69.35	4.76	0.80	0.80	0.00
2015	14.30	15.99	11.83	64.60	67.45	4.41	0.80	0.80	0.00
2016	14.50	15.80	8.97	64.40	65.66	1.96	0.80	0.80	0.00
2017	16.70	15.63	6.41	62.70	63.98	2.04	0.80	0.80	0.00
2018	16.50	15.47	6.22	61.60	62.40	1.29	0.80	0.80	0.00
2019	16.90	15.33	9.27	58.90	60.90	3.39	0.90	0.90	0.00
2020	15.20	15.21	0.04	57.90	59.48	2.72	0.80	0.90	12.50
2021	15.30	15.09	1.38	65.00	58.13	10.57	0.90	0.90	0.00

sensitivity. The sensitivity calculation formula (Wang *et al.*, 2022) is as follows:

$$\begin{cases} S_{ij} = (\Delta D_{(i,j)}/D_{(i,j)})/(\Delta X_{(i,j)}/X_{(i,j)}) \\ S_i = (1/n) \times \sum_{j=1}^{n} S_{ij} \end{cases} \tag{8.2}$$

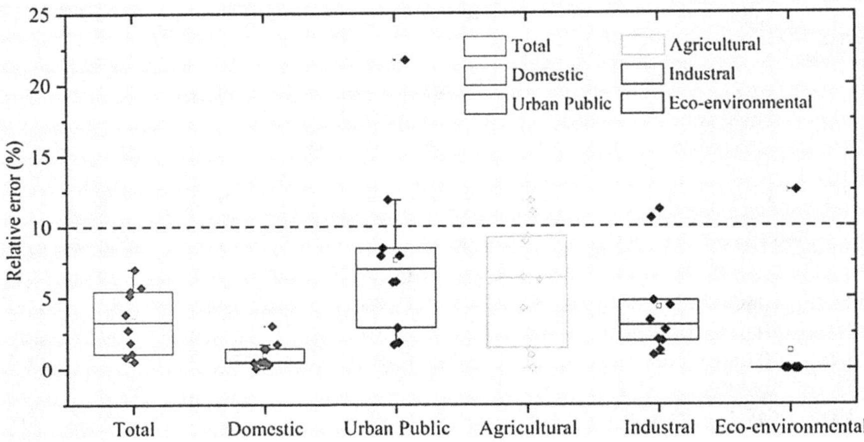

Figure 8.3. Validation of the system dynamics model results for water demand prediction in Shanghai.

where, S_{ij} denotes the sensitivity of the ith indicator to the water demand in the jth year, while S_i is the average sensitivity of indicator i. $X_{(i,j)}$ and $D_{(i,j)}$ represent the value of the ith indicator and its corresponding water demand in the jth year. Changes in these values are denoted by $\Delta X_{(i,j)}$ and $\Delta D_{(i,j)}$, respectively.

Using Eq. (8.2), the sensitivity of eight key factors influencing water demand was computed (Table 8.3). The growth rate of the effective irrigated area shows a negative sensitivity, indicating an inverse relationship with water demand. In contrast, other factors exhibit positive sensitivities. Notably, the two indicators related to industry (I_1 and I_2), with sensitivities of 0.510 and 0.445, respectively, are identified as having the greatest impact on water demand. This underscores the need for increased technological investment to improve water use efficiency alongside industrial growth.

(4) **Development mode settings**

Five modes were developed to predict Shanghai's future water demand, considering the city's development objectives and indicator sensitivities (Table 8.3):

Mode 1 (M1): Business as usual extends historical socio-economic growth trends into the future.

Table 8.3. Average sensitivity and development mode settings of indicators influencing water demand.

Indicator	Average sensitivity	M1		M2		M3		M4		M5	
		2035	2050	2035	2050	2035	2050	2035	2050	2035	2050
I_1	0.510	29.14	17.68	29.14	17.68	29.14	17.68	29.14	17.68	26.23	15.91
I_2	0.445	2.26	1.72	4.98	3.78	2.26	1.72	4.98	3.78	4.98	3.78
I_3	0.237	66.63	77.34	66.63	77.34	66.63	77.34	66.63	77.34	63.30	73.47
I_4	0.121	466.72	462.39	466.72	462.39	466.72	462.39	466.72	462.39	443.39	439.27
I_5	0.115	5.37	3.87	5.91	4.26	5.37	3.87	5.91	4.26	5.91	4.26
I_6	0.059	0.80	0.17	0.80	0.17	0.80	0.17	0.80	0.17	0.76	0.16
I_7	−0.045	−0.97	−0.79	−0.97	−0.79	−0.97	−0.79	−0.97	−0.79	−0.97	−0.79
I_8	0.019	1.83	1.78	1.83	1.78	7.33	7.12	7.33	7.12	7.33	7.12

Note: The indicators I_1–I_8 respectively indicate water demand per 10,000 RMB of industrial added value (m^3 /10^4 RMB), industrial added value growth rate (%), per capita domestic water demand (m^3/person), water demand per mu for irrigation (m^3/mu), tertiary industry added value growth rate (%), water demand per 10,000 RMB of tertiary industry added value (m^3/10^4 RMB), effective irrigated area growth rate (%), and population growth rate (‰).

Mode 2 (M2): Economic acceleration development is based on Shanghai's *Three-Year Action Plan for High-Quality Manufacturing Development (2023–2025)*, which anticipates industrial added value exceeding 1.3 trillion RMB by 2025. To achieve this target, the growth rate of industrial added value is increased in this mode based on M1. Additionally, the growth rate of tertiary industrial added value is projected to increase by 10% after 2025. Except for these economic adjustments, all other indicators follow historical trends.

Mode 3 (M3): Population acceleration development addresses aging demographics. With over one-third of Shanghai's population over 60 as of 2022, boosting birth rates and attracting younger demographics are necessary. This mode introduces a yearly increase of 200,000 people, beginning in 2035, guided by the *Shanghai Urban Master Plan (2017–2035)* and Shanghai's demographic study by Wang *et al.* (2015). Under this mode, economic and other parameters are consistent with historical trends, as in M1.

Mode 4 (M4): Socio-economic acceleration development combines the parameter settings of M2 and M3.

Mode 5 (M5): Comprehensive development aims to enhance water use efficiency. Considering the model's sensitivity, it assumes a 10% decrease in industrial efficiency and a 5% increase in tertiary, irrigation, and domestic sectors.

8.2.3.2 *Available water calculation*

To calculate the available water in Shanghai's reservoirs, we used actual salinity measurements at the water intake points during the 2022 drought (Table 8.4). Based on the start and end times of saltwater intrusion in the Chenhang, Qingcaosha, and Dongfengxisha reservoirs within the Yangtze Estuary, we determined whether each reservoir could draw water. During saltwater periods, the chloride concentration in the water exceeds 250 mg/L (salinity is 0.45 PSU), rendering it unsuitable for tap water production. Consequently, the reservoir cannot be drawn upon during those periods. However, during the intervals between two salty tides, when the chloride concentration drops below 250 mg/L, emergency freshwater replenishment can be performed by pumping water from rivers (Li *et al.*, 2009).

Table 8.4. Measured salinity at the intake of estuarine reservoirs from September 1 to December 8, 2022 (from Shanghai Water Bureau).

Reservoir	Start time	End time	Duration	Maximum c(Cl^-) at intake
Chenhang	9.14	9.20	Five days and 12 hours	2166 mg/L
	9.21	10.18	26 days and 12 hours	
	10.22	continue	37 days (as of 8 Dec.)	
Qingcaosha	9.5	continue	94 days (as of 8 Dec.)	3959 mg/L
Dongfengxisha	9.1	9.4	Three days and five hours	2176 mg/L
	9.6	9.19	13 days	
	9.20	10.18	27 days and 15 hours	
	10.23	continue	39 days (as of 8 Dec.)	

Pumped water after meeting regional demand is stored, or, if insufficient, supplemented by the reservoirs (Tu *et al.*, 2022). The Jinze Reservoir, unaffected by salt tides but impacted by drought-induced low levels, relies on pumped intake. Since cumulative precipitation and evaporation in Shanghai from September to December 2022 (the months this study primarily focuses on) were almost equal—327.2 mm and 333.7 mm, respectively—we did not consider them in this analysis.

The available water during salt tides is calculated as follows:

$$\mathrm{RAW} = \sum \mathrm{RAW}_i \tag{8.3}$$

$$\mathrm{RAW}_i = \mathrm{RAW}_{0i} + \sum \mathrm{PY}_{ai} - \sum RWS_{ai} \tag{8.4}$$

where RAW is the total available water in the reservoirs, RAW_i and RAW_{0i} are the available water and the initially available water of reservoir i, PY_{ai} is the actual water pump yield, and RWS_{ai} is the water supply for reservoir i.

Sediment deposition in the reservoir area reduces capacity. Given the Yangtze River's high sediment content, the dead storage capacity of the reservoir is not taken into account. The initial storage volume of a reservoir plays a crucial role in determining supply duration during salt tide periods. For this study, we considered three initial water levels of the Qingcaosha Reservoir—design high, operational high, and average—to estimate available supply. The Qingcaosha Reservoir was selected as the specific research focus due to its substantial capacity for storage and water supply, which are pivotal for ensuring Shanghai's WSS. Its effective capacity and daily water supply capacity exceed those of other reservoirs by 43–47 times and 2–30 times, respectively (Figure 8.1(c)). For the Chenhang, Dongfengxisha, and Jinze reservoirs, we assume operations at the highest water level, equating the initial supply with effective storage.

The theoretical and actual pump yields for the reservoirs are calculated during the salt tide period as follows:

$$\mathrm{PY}_{ti} = V_{pi} \times T \tag{8.5}$$

$$\mathrm{PY}_{ai} = \begin{cases} \mathrm{ES}_i - \mathrm{RAW}_{d0i} + RWS_{ai}, \mathrm{RAW}_{d0i} + \mathrm{PY}_{ti} > \mathrm{ES}_i \\ \mathrm{PY}_{ti}, \mathrm{RAW}_{d0i} + \mathrm{PY}_{ti} \leq \mathrm{ES}_i \end{cases} \tag{8.6}$$

where PY_{ti} and PY_{ai} are the theoretical and actual pump yields for reservoir i, respectively; V_{pi} is the pump intake rate for reservoir i; T is the duration of pumping; ES_i is the effective storage of reservoir i; and RAW_{d0i} is the initially available water of reservoir i for the day.

The theoretical daily water supply is calculated based on the reservoir's daily water demand, taking into account losses as described in Eq. (8.7). Reservoir water primarily supports domestic use, urban public needs, general industry (comprising about 15% of total industrial demand), and the eco-environment. This demand, predicted by the SD model, is divided by 365 to determine daily requirements, as shown in Eq. (8.8). Leveraging Jinze Reservoir's supply capacity, which is unaffected by saltwater intrusion, is crucial to reducing pressure on the Qingcaosha Reservoir. Given its limited output, the Dongfengxisha Reservoir is expected to function at full capacity. The allocation from the Qingcaosha and Chenhang reservoirs is calculated after subtracting the contributions from Jinze and Dongfengxisha, and water distribution is adjusted based on the actual supply ratios observed over the past five years:

$$RWS_{dt} = RWD_d \times (1 + R_{wl}) \tag{8.7}$$

$$RWD_d = RWD_a/365 \tag{8.8}$$

where RWS_{dt} represents the theoretical daily water supply from the reservoir; RWD_a and RWD_d are the annual and daily reservoir water demands, respectively; and R_{wl} is the water loss rate, which is set at 7% in this study (Liu *et al.*, 2009).

8.2.3.3 *WSS risk assessment*

Utilizing the SD method, we modeled freshwater supply–demand balance analysis during saltwater intrusion periods (Figure 8.2(c)). The model covered Shanghai's four major reservoirs from September 1, 2022, to January 30, 2023, spanning 153 days, with a daily time step. The simulation accounted for daily water intake and water supply in each reservoir and analyzed the balance of water supply and demand. Three indexes, i.e., WSI, DSA, and WSA, were used to assess risks.

WSI, the ratio of water use to available supply, is a recognized index of water scarcity (He *et al.*, 2021; Raskin *et al.*, 1997). In this study, WSI is determined by the ratio of theoretical reservoir water supply to reservoir availability (Eq. (8.9)). It mainly focuses on the total available water without considering actual reservoir supply capacity, making it suitable for assessing annual WSS risks:

$$\mathrm{WSI}_j = RWS_{tj}/\mathrm{RAW}_j \tag{8.9}$$

where WSI_j is the water stress index at time j; RWS_{tj} is the theoretical reservoir water supply at time j, the sum of the reservoir water demand and the water lost; and RAW_j is the reservoir available water at time j. A higher WSI indicates greater risks, with a WSI > 1 indicating unmet demand. A WSI ≤ 1 means the amount of reservoir available water is sufficient to meet the demand, while determining whether a water shortage exists requires considering the water supply capacity of the reservoirs. If the supply capacity cannot meet the demand, a water shortage will occur.

A shortage occurs when the actual daily supply from reservoirs falls below demand plus losses, and its calculation formula is given by Eq. (8.10). The duration until a shortage is termed DSA, with fewer days and greater shortage amounts indicating increased risk. These two indexes, considering the reservoir's actual supply capacity, are suitable for assessing intra-year WSS risks:

$$WS_{dj} = RWS_{aj} - RWS_{tj}(RWS_{aj} < RWS_{tj}) \tag{8.10}$$

$$WS = \sum WS_{dj} \tag{8.11}$$

where WS_{dj} and WS are the water shortage amount at day j and the total water shortage amount, respectively. RWS_{aj} is the actual reservoir water supply at time j.

8.2.3.4 *DAPP approach*

The DAPP method offers a systematic framework for addressing the uncertainties associated with climate change. By outlining a series of potential future actions, DAPP enables policymakers to plan and adjust policies proactively, ensuring adaptability and flexibility. Its essence lies in identifying "tipping points," where existing current management approaches fail to meet objectives, requiring new or supplementary strategies. Detailed guidance on applying DAPP and pinpointing tipping points can be found in studies by Haasnoot *et al.* (2013) and Babaeian *et al.* (2023).

In this research, we employed an SD model alongside various socio-economic development modes and changes in reservoir water levels to assess the efficacy of different adaptation strategies. By simulating future variations in WSI and potential WSA, we identified critical "tipping points"—moments when the actual water supply cannot meet the demand, indicating that the current strategy can no longer ensure WSS. Integrating these analyses, we developed a DAPP map that visually outlines potential actions and their timing across different decision points.

8.3 Results

8.3.1 *Water demand trends in Shanghai*

8.3.1.1 *Water use patterns for 2000–2021*

Between 2000 and 2021, Shanghai's water use patterns exhibited two distinct phases: an increase from 2000 to 2010, followed by a gradual decline from 2011 to 2021 (Figure 8.4(a)). During the growth phase, annual water use rose from 10.84 to 12.63 billion m^3 in 2010. This period saw increases in agricultural, industrial, domestic, and urban public water use, with increases of 177, 614, 597, and 396 million m^3, respectively. Conversely, water use in the eco-environments dropped from 0.23 billion m^3 in 2004 to 0.12 billion m^3 in 2010.

In contrast, from 2011 to 2021, the overall water use trended downward due to improvements in water efficiency, dropping to 10.58 billion m^3 in 2021, a reduction of 2.05 billion m^3 from 2010. Water use declined in agriculture, industry, urban public services,

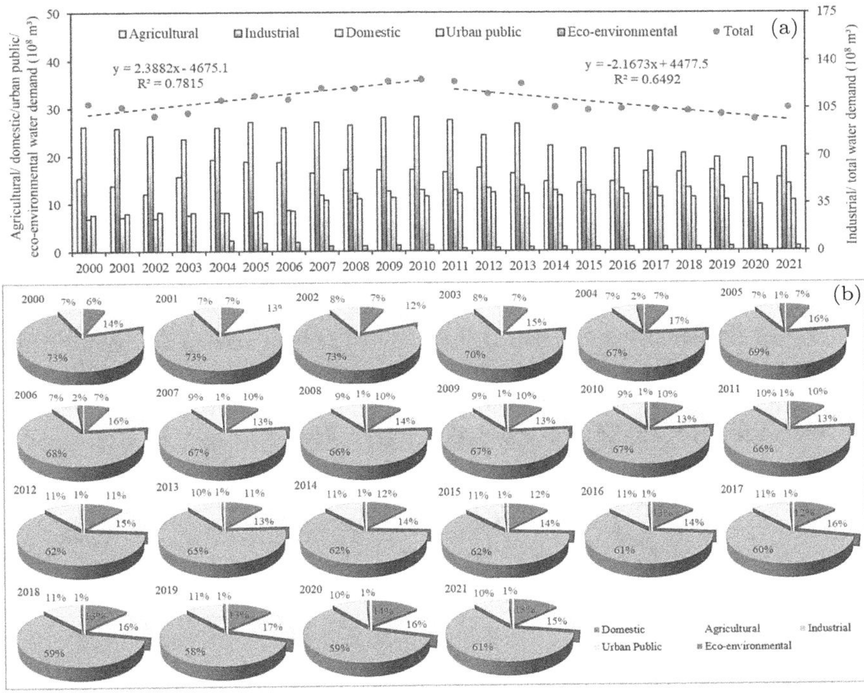

Figure 8.4. Trends in water usage (a) and composition (b) in Shanghai from 2000 to 2021.

and eco-environment to 1.53, 6.5, 1.06, and 0.09 billion m^3 in 2021, respectively, with industrial water use seeing the largest reduction (23.35%). Domestic water use, however, increased by 9.70% to 1.40 billion m^3. Additionally, the COVID-19 pandemic in 2020 led to significant reductions in societal production activities, causing a notable drop in total water use (down by 3.37% from 2019). However, with the resumption of normal activities in 2021, industrial and urban public water use rose, leading to a significant increase in total water use by 8.51% from 2020.

Over the past 20 years, Shanghai's water consumption composition has remained fairly consistent (Figure 8.4(b)). In 2000, industry dominated at 73%, followed by agriculture at 14%, urban public services at 7%, and domestic use at 6%. A shift occurred after 2007, with domestic use overtaking urban public services, ascending to the third-largest category. By 2021, the distribution had adjusted to 61% for industrial, 15% for agricultural, 13% for urban public, 10% for domestic, and 1% for eco-environmental. This change signifies a noteworthy reduction of 12% in industrial use, alongside increases of 7% in domestic and 3% in urban public use. Agricultural and eco-environmental proportions remained largely stable, highlighting the city's targeted efforts in water resource management and conservation.

8.3.1.2 *Water demand prediction for 2022–2050*

Shanghai's future water demand will diverge according to five development modes. Under modes emphasizing economic growth (M2, M4, and M5), demand is projected to rise, peaking between 10.85 and 11.80 billion m^3 by 2050 (Figure 8.5(a)). In contrast, modes based

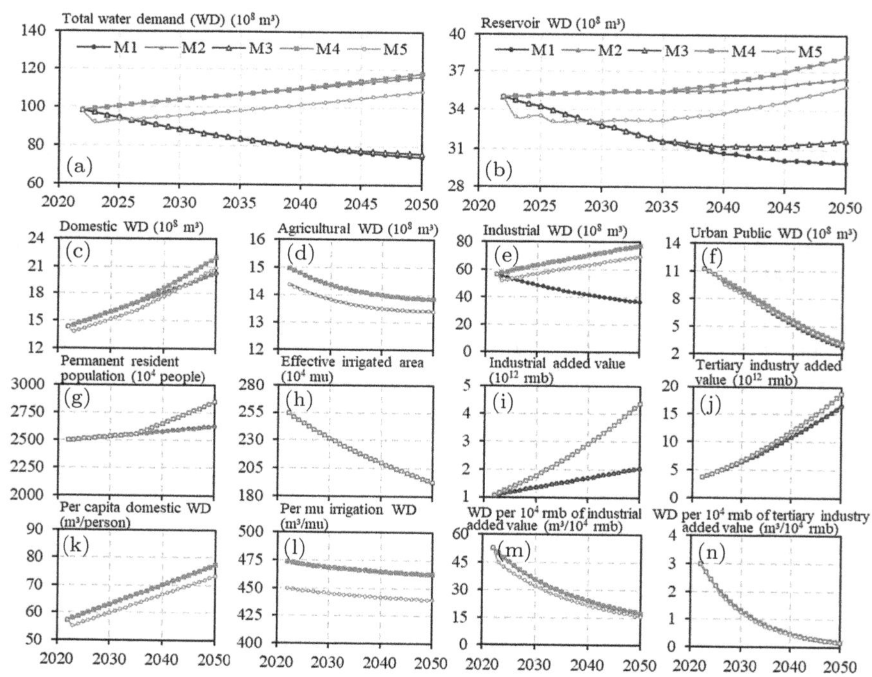

Figure 8.5. Water demand (WD) prediction in Shanghai (2022–2050) using a system dynamics model across five development modes: (a)–(b) total and reservoir-specific demand trends; (c)–(n) sectoral demand, efficiency, and scale. Tertiary industry added value includes construction.

on historical trends (M1) and population growth (M3) project a gradual decrease, bottoming out at 7.48–7.65 billion m^3 by 2050. M1 predicts a consistent reduction in reservoir water demand, whereas M2 and M4 indicate an uptick, and M3 and M5 suggest an initial drop followed by a rise. Notably, M5 shows a brief decrease before increasing demand, while M3 experiences a more extended period of decline. By 2035 and 2050, reservoir water demand is expected to reach 3.16–3.54 billion m^3 and 3–3.83 billion m^3, respectively (Figure 8.5(b)).

Domestic water demand is expected to rise in all scenarios, with M3 and M4 showing a higher population growth rate, potentially reaching 2.20 billion m^3 by 2050. Shanghai's population is projected to align with the city's master plan, capping at nearly 25 million by 2035 and, depending on growth rates, is forecasted to range from 26.25 million to 28.47 million by 2050 (Figure 8.5(g)). Per capita domestic water demand is also expected to climb, with a range of 73.47–77.34 m^3/person by 2050, depending on whether water efficiency is high or low (Figure 8.5(k)).

Agricultural demand is predicted to decline due to a shrinking effective irrigation area and per-unit irrigation water demand, with projected demands of 1.34–1.39 billion m^3 by 2050 across modes (Figures 8.5(d), 8.5(h), and 8.5(l)). Ecological water demand is modest, forecasted at 0.11 billion m^3 in 2035, growing slightly to 0.14 billion m^3 by 2050.

Industrial water demand is set to increase under economy-focused modes (M2, M4, and M5) and decrease under historical and population-centric modes (M1 and M3), with projections for 2035 and 2050 of 4.47–6.72 billion m^3 and 3.64–7.75 billion m^3, respectively (Figure 8.5(e)). Industrial added value is poised to rise, reaching 2.31 and 4.38 trillion RMB by 2035 and 2050 under aggressive economic modes, exceeding less growth-focused modes by 0.78 and 2.32 trillion RMB (Figure 8.5(i)). Industrial efficiency is anticipated to improve over time, with 26.23–29.14 $m^3/(10^4$ RMB by 2035 and 15.91–17.68 $m^3/$ $(10^4$ RMB by 2050 (Figure 8.5(m)). Despite the expansion of the tertiary sector, urban public water demand is anticipated to diminish due to enhanced efficiency, dropping from 0.68–0.72 billion m^3 to 0.29–0.32 billion m^3 by 2050 (Figures 8.5(f), 8.5(j), and 8.5(n)).

8.3.2 *Drought impacts on Shanghai's water supply*

Historically, Shanghai faced severe water security challenges during the extreme droughts of 1978, 2006, and 2022 in the Yangtze River Basin. The 1978 drought led to salinity intrusion into the Yangtze Estuary and even the Huangpu River, surrounding Chongming Island with saltwater for nearly 100 days. In October 2006, a salt tide reduced the daily water supply in Pudong New Area by a quarter. The 2022 drought, intensified by the impact of typhoons, led to salinity intrusion severity that surpassed that of the previous two events.

In 2022, the Datong station recorded unusual discharge patterns. In June, discharge peaked at 58,200 m^3/s—significantly higher than in 2018 (a typical normal year)—and then sharply declined to levels below those of the 2006 extreme drought by July (Figure 8.6(a)). From September to mid-February, discharge remained below the saltwater intrusion threshold of 15,000 m^3/s for an unprecedented

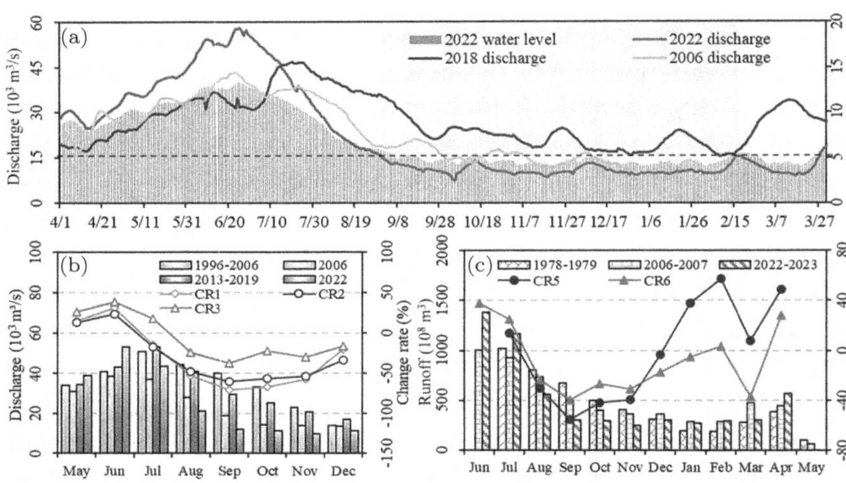

Figure 8.6. Discharge and runoff at Datong station during multi-year, normal, and extreme drought years. (a) Daily discharge and water level during the drought years 2006 and 2022 and the normal year 2018. Discharge data from January to March 2007 is unavailable. (b) 2022 monthly discharges against long-term averages and 2006. CR1, CR2, and CR3 represent changes in 2022 compared to 1996–2006 and 2013–2019 (before and after the Three Gorges Reservoir) and 2006, respectively. (c) Runoff comparisons from the 1978, 2006, and 2022 droughts. CR5 and CR6 are the changes in 2022 compared to 1978 and 2006, respectively.

168 days. The water discharge at Datong was below 15,000 m^3/s, despite an early October release of water from the TGR. Monthly discharge at Datong from July to October was down by 17.83–60% during 2013–2019 (Figure 8.6(b)). Runoff was significantly lower than historical droughts of 1978 and 2006, from August to January, with decreases of 3.47–55.19% compared to 1978 and 5.27–39.25% compared to 2006 (Figure 8.6(c)).

Saltwater intrusion, usually occurring from November to March, struck the Yangtze Estuary unusually early in the summer of 2022, exacerbated by basin-wide drought and typhoons. Chloride levels at Sanjia Port and other locations in Pudong soared past safe thresholds from August 10–12. In September, repeated salinity peaks at Qingcaosha and Chenhang reservoirs indicated a heightened intrusion pattern (Table 8.4). By December 8, these reservoirs experienced extended periods of inoperative intake—37, 94, and 39 days—due to high chloride levels, which peaked at 2,166–3,959 mg/L, well above capacity limits. With river discharge at Datong station remaining under 15,000 m^3/s, persistent low water levels predicted persistent saltwater intrusion, halting reservoir intakes. Therefore, the risk analysis of WSS in Section 8.3.3 will focus on the period from September 1 to January 30 of the following year.

8.3.3 *Shanghai WSS risks under normalized extreme climate conditions*

For the WSS risk analysis under normalized extreme climate conditions in Shanghai, the study modeled 15 scenarios, combining three water level states of the Qingcaosha Reservoir with five development modes (Figure 8.7). The findings demonstrate that higher initial water levels correspond to lower risks. In 2035, across varying initial water states, the average WSI will be 1.47–1.82, the average DSA will range from 40 to 75, and the average WSA will be 490–684 million m^3. By 2050, risks will escalate, with the average WSI rising to 1.51–1.87, DSA decreasing slightly, and WSA rising to 534–730 million m^3. Within each development mode, given consistent water levels, WSIs follow a specific pattern: They decrease in historical trends (Si-1, $i = 1, 2$, and 3), rise with economic acceleration (Si-2 and Si-4), and initially dip before rising with population acceleration (Si-3) and comprehensive development (Si-5). The highest-risk

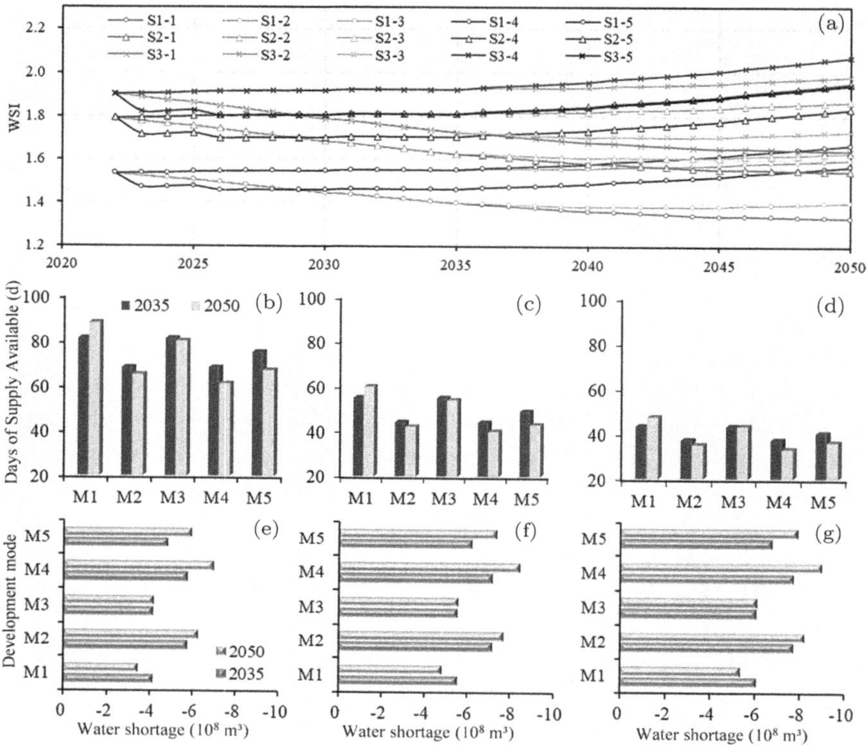

Figure 8.7. Shanghai's water supply security risks under drought conditions from 2022–2050 across 15 scenarios: (a) water stress index (WSI) variations; (b)–(d) days of supply available from reservoirs for different initial water levels (design high, operational high, and average) of the Qingcaosha Reservoir in 2035 and 2050; (e)–(g) water shortage amount projections corresponding to (b)–(d).

scenario by 2050 projects a WSI of 2.07, with only 33 supply days and a deficit of 896 million m^3, while the least-risk scenario anticipates a WSI of 1.33, with 88 supply days and a 343 million m^3 deficit.

Taking the scenario S2-5 (operational high water level with comprehensive development mode) as an example, we further explored Shanghai's WSS risks in 2022, 2035, and 2050 (Figure 8.8). In 2022, with a water supply of 10.25 million m^3/day, reservoirs could support 45 days of normal water distribution. During this period, 461 million m^3 of water is supplied, with 268 million m^3 pumped. By day 46, a shortfall of 3.28 million m^3 occurs, and the WSI reaches 0.96. Despite emergency pumping of the Chenhang and Dongfengxisha

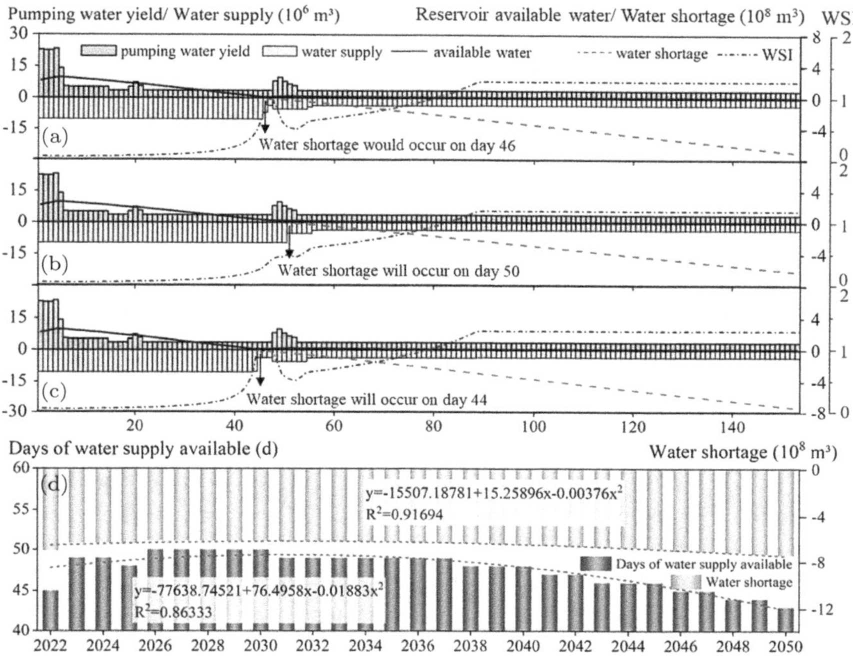

Figure 8.8. Water supply security risks for Shanghai during drought conditions under scenario S2-5 from 2022 to 2050: (a)–(c) available water and WSI variations in 2022, 2035, and 2050, respectively; (d) changes in days of supply availability and water shortage amounts.

reservoirs, reducing the WSI to below 1, shortages persist due to deficits at the Qingcaosha Reservoir and other sources failing to meet demand. By day 56, the daily deficit escalates to 6.5 million m^3, resulting in a total shortfall of 702 million m^3 by the end of January (Figure 8.8(a)).

In 2035, daily water demand will slightly decrease, and the DSA will extend by four days compared to 2022. WSA will commence on day 50 with a WSI of 0.49, subsequently rising. Despite overall water availability exceeding demand, specific limitations at individual reservoirs resulted in a supply shortage. By day 57, the daily deficit will reach 5.97 million m^3, leading to a total deficit of 623 million m^3 by January (Figure 8.8(b)). In 2050, the theoretical daily water supply will increase to 10.50 million m^3, and DSA will be shortened by six days compared to 2035. The deficit will occur on day 44, with a WSI of 0.94, and the daily gap will widen to 6.75 million m^3, summing up to 739 million m^3 by January (Figure 8.8(c)).

From 2022 to 2050, under scenario S2-5, the duration of supply follows a parabolic trend, initially increasing and then decreasing, with a corresponding trend in WSA first decreasing and then increasing. The water supply gap persists, with 2026 presenting the least risk to WSS (Figure 8.8(d)).

8.4 WSS Response Strategies

8.4.1 *Enhance water use efficiency and control demand*

Since 2010, Shanghai's water consumption has been on a declining trend (Figure 8.4(a)). Wang *et al.* (2023) reported that actual water use is influenced by regional resources and policies, while water demand represents an ideal water volume. The reduction in water use can be attributed to improved water-saving efficiency, statistical standards adjustment, measurement errors, and water scarcity. Following China's 2012 water management reforms, Shanghai established a management framework by 2014 as a pilot program, significantly improving water use efficiency (SWA, 2012). Our study aimed to align statistical standards across indicators, referencing *Shanghai Water Resource Bulletins* and national statistics. Local and transit water sources exceeded the long-term average by 44.15% and 2.95% (Table 8.1), suggesting that available supply was not the issue for the reduction. Despite events such as the COVID-19 pandemic abruptly affecting usage, the overall trend is expected to normalize (Liu *et al.*, 2022). Thus, our SD model for predicting water demand is based on an analysis of Shanghai's water use under strict water management.

Policy adjustments directly influence future water demand. Many developed countries reached their peak water use in the 1980s and 1990s, followed by "inverted-V" changes, stabilization, or multiple peaks (Wang *et al.*, 2023). Sensitivity analysis of influencing factors can help guide water management strategies. Industrial water demand, which accounts for over half of Shanghai's total demand, significantly influences the overall trend. Our findings show that accelerated economic growth increases demand and risks, while efficiency gains, such as those in the comprehensive development scenario (M5), could mitigate these risks. For example, increased efficiency could reduce total water demand by 945 million tons and daily reservoir demand by 660 thousand m^3 by 2050 (Figure 8.5). While an increased population growth rate has a minor effect on total water demand, it significantly pressures reservoir supplies, and this pressure is expected to grow over time. This is because domestic water demand accounts for a relatively small proportion of total water demand, yet it is the primary source of water for Shanghai's reservoirs. Therefore, Shanghai should continue to improve water use efficiency and manage population growth amidst the new climate extremes norm.

8.4.2 *Strengthen monitoring and information sharing*

The inflows to the Yangtze River Basin, influenced by climate and human activities, directly affect estuarine freshwater conditions. Particularly during droughts, reduced runoff intensifies saltwater intrusion, shrinking freshwater availability. This scenario is further complicated by climate change, which is anticipated to increase both the frequency and severity of such droughts in the Yangtze River Basin (Yue, 2022).

Sediment accumulation in reservoirs, such as the TGR, has accumulated over 2 billion tons of sediment since its inception, reducing storage capacity and reliability (Yuan *et al.*, 2023). The Qingcaosha Reservoir also suffers from a similar issue, which hinders water flow and storage (Yan and Wu, 2015). Climate change exacerbates these problems by leading to more frequent and intense flooding, which in turn accelerates sediment accumulation, thereby necessitating a focus on the sustainability of these resources (Fan *et al.*, 2023; Yao *et al.*, 2023).

Beyond climate factors, human activities in the middle and lower Yangtze River add uncertainty to estuarine inflows. The TGR usually stores water from September to October and gradually releases it from April to June (Figures 8.9(a) and 8.9(b)). Saltwater intrusion in 2022 threatened Shanghai's water supply, prompting emergency releases from the reservoir in early October. The TGR's capability to release water is limited by the need to maintain its comprehensive benefits at a low water level. Currently, 117 large upstream reservoirs with a total storage capacity of 84.2 billion m^3 play critical

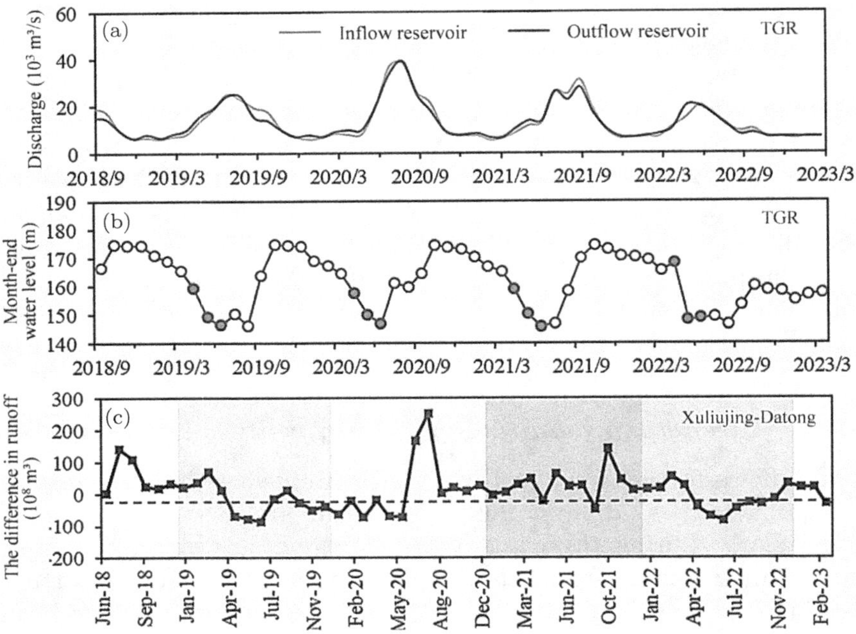

Figure 8.9. River discharge and runoff variations at the Three Gorges Reservoir (TGR), Datong, and Xuliujing from 2018 to 2023: (a) TGR's monthly inflow and outflow; (b) TGR's monthly water level, with red circles representing April to June each year; (c) the monthly runoff gap between the Xuliujing and Datong stations.

roles in water supply, power generation, and flood control, yet their drought resilience needs enhancement. Additionally, significant water diversion and withdrawal activities between the Datong and Xuliujing stations, including the South-to-North Diversion Project, impact the flow, exacerbating estuarine saltwater intrusion during the 2022 drought (Figure 8.9(c)).

To address water inflow uncertainties in the Yangtze River Basin, it is essential to improve water monitoring downstream of Datong and to better utilize the Xuliujing station. Strengthening sediment control in reservoirs, boosting upstream dams' drought resilience, and regulating water intake during low-flow periods are critical. Leveraging smart platforms for data sharing and integration is key to promptly identifying and mitigating water supply risks in Shanghai in the face of normalized extreme climate conditions.

8.4.3 *Lessons from mega-cities' water systems*

8.4.3.1 *Building new reservoirs*

Building reservoir networks ensures a stable water supply for mega estuarine cities. A survey of seven major global estuarine cities (New York, Greater Vancouver, London, Paris, Greater Sydney, Tokyo, and Shanghai) reveals that most have over ten reservoirs to maintain a stable supply and quality of water. Sydney has the highest per capita reservoir capacity, followed by New York, while Shanghai ranks third from last, with only 10% of New York's per capita capacity (Figure 8.10 and Table 8.5). Notably, the bottom-ranked London and Paris source 30–55% of their drinking water from groundwater, which differs from Shanghai's reliance on surface water.

The Qingcaosha Reservoir provides 54% of Shanghai's water supply. Although this capacity is generally adequate under normal conditions, delays in storage or pollution during droughts pose serious threats. Thus, Shanghai is advised to increase the number of reservoirs to better utilize transit water, potentially by constructing a chain of Yangtze Estuary reservoirs (Jin *et al.*, 2004). Additionally, leveraging local resources by building small reservoirs or storage ponds in areas with good water quality for storing rainwater, river water, or other sources can provide support during droughts or other crises.

8.4.3.2 *Integrated water transfer in the Yangtze river delta*

Diversifying water sources reduces supply risks. For instance, New York's water comes from the Catskill, Delaware, and Croton watersheds. Tokyo sources from the Tone, Arakawa, and Tama river systems, and Greater Sydney from five interregional sources. Shanghai, however, relies on the Yangtze Estuary and the Huangpu River. Given Shanghai's downstream location, optimizing water resources from a regional perspective, including the integrated Yangtze River Delta, is crucial to guard against the increasingly unpredictable climate extremes. Consideration could be given to transferring water from water-rich and quality areas such as western Zhejiang and southern Anhui. Additionally, enhancing emergency water supply plans and strengthening inter-source cooperation is recommended (Chen *et al.*, 2018), as well as promoting the interconnectivity of

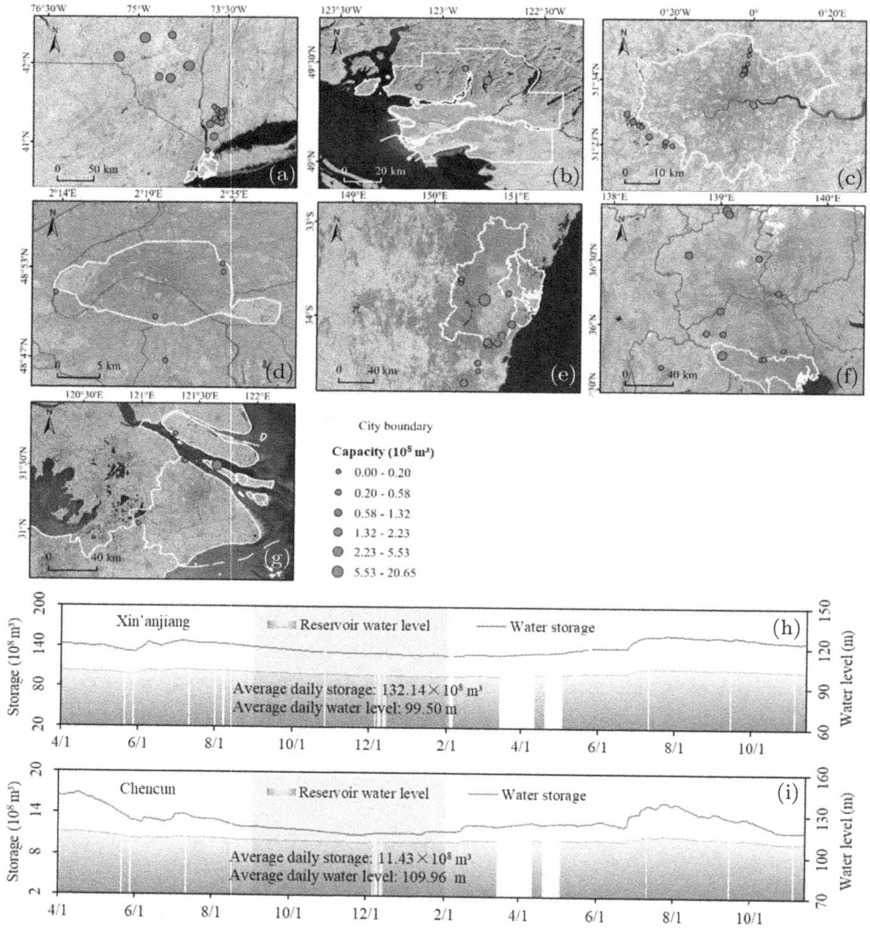

Figure 8.10. Reservoir distribution in major international estuarine megacities and water storage in the Xin'anjiang and Chencun reservoirs during 2022–2023. (a)–(g) are New York, Greater Vancouver, London, Paris, Greater Sydney, Tokyo, and Shanghai, respectively. The shaded areas in (h) and (i) represent September 2022 to January 2023, the period of drought conditions studied, and blank areas indicate missing data.

water pipelines between water plants to improve the level of emergency support.

If a drought similar to the one in 2022 recurs in 2050, Shanghai could face critical challenges, with DSA potentially reducing to

Table 8.5. Reservoir statistics of major international megacities.

City	Population (10^4)	Reservoir	Capacity (10^8 m^3)	Per Capita Storage (m^3)
New York	834	20	21.92	234.69
Metro Vancouver	256	3	3.13	122.27
London	898	22	2.08	23.16
Paris	216	5	0.01	0.46
Greater Sydney	486	15	27.01	555.76
Tokyo Metropolis	1407	13	8.92	63.40
Shanghai	2475	5	5.93	23.96

only 33 days and WSA reaching 896 million m^3, necessitating an additional daily water supply of 6 million m^3. The Yangtze River Delta, excluding Shanghai, has 58 large reservoirs with a storage capacity above 100 million m^3. In 2021, the Anhui and Zhejiang provinces had water resources of 88.33 and 134.47 billion m^3, respectively, with unutilized resources offering potential relief for Shanghai. Xin'anjiang and Chencun, the largest reservoirs in Zhejiang and Anhui, respectively, are located on the Qiantang River and Qingyi River (a lower Yangtze tributary). Their capacities are 21.63 and 2.69 billion m^3, respectively, presenting different opportunities for aiding Shanghai, with Xin'anjiang Reservoir offering greater potential due to its larger surplus (Figure 8.10).

8.4.4 *DAPP for Shanghai's WSS risks*

Based on SD simulations and feasible solutions discussed earlier, a DAPP map was created to guide Shanghai's water supply strategies from 2022 to 2050 under the highest-risk scenario (S3-4) (Figure 8.11). This map illustrates diversified water security strategies Shanghai might employ amid compounding pressures of socio-economic development, saltwater intrusion, and critically low Qingcaosha Reservoir levels. Strategies include improving water use efficiency, adjusting reservoir operational levels, implementing emergency water measures, constructing new reservoirs, and water transfers, with implementation costs increasing with complexity.

The DAPP map indicates Shanghai may need to timely adjust its policies promptly to manage water security challenges. If existing

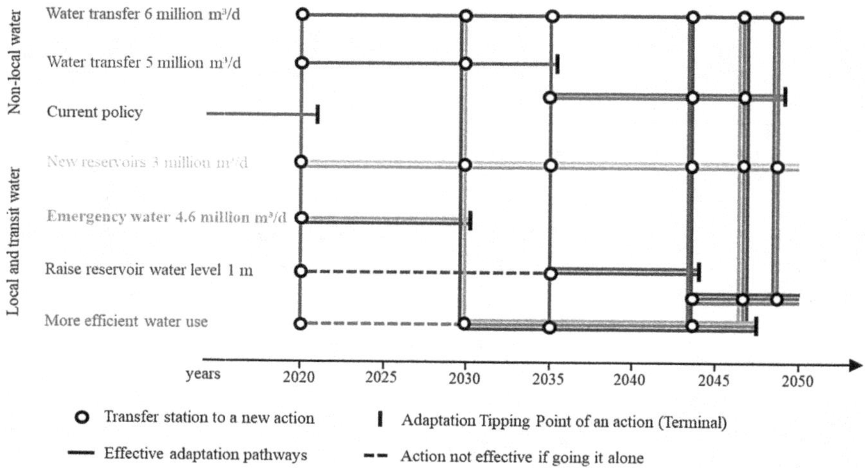

Figure 8.11. The map of dynamic adaptive policy pathways for water supply security in Shanghai (2022–2050) under the highest risk mode (S3-4). The font color corresponds to the color of the lines. Double or triple lines indicate the simultaneous implementation of two or three measures.

strategies fall short, alternative paths can be pursued at predetermined decision points, with dual and triple lines denoting strategies implementing two or three measures simultaneously.

To ensure Shanghai's future water security, the focus should be on sustainable local water resource management and the full utilization of transit resources, such as the Yangtze River. Although improving water use efficiency or adjusting reservoir water levels are cost-effective, these approaches alone are insufficient to withstand the challenges of extreme climate events and must be integrated with other strategies. In the event of similar extreme events, leveraging local emergency resources (e.g., the Songpu Bridge water source, emergency intakes, or micro-storage facilities), alongside efficiency improvements and maintaining reservoir levels, can secure short-to-mid-term water supply. For the long term, constructing new reservoirs and water transfers may be necessary. However, if local emergency supplies are unreliable during extreme events, costlier options such as new reservoir construction or water transfer must be considered. A multi-source strategy that incorporates local, transit, and inter-regional resources enhances urban resilience against extreme weather and climate change while reducing reliance on a single water source. Especially after 2035, multiple measures should be implemented concurrently to strengthen water resource management.

Shanghai's future water strategies need to find a balance within ongoing assessments and dynamic adjustments to ensure long-term water security and sustainability. Employing dynamic management addresses the uncertain future, ensuring the continuity and adaptability of Shanghai's water supply system.

8.5 Conclusions

This study assessed Shanghai's water security risks amid normalized climate change by combining the application of SD models and the DAPP method, considering 15 scenarios, including economy, population, water efficiency, and reservoir levels, and offering targeted strategies. The key conclusions are as follows:

(1) From 2000 to 2021, Shanghai's water use experienced an increase until 2010, followed by a decline, indicating a shift toward more efficient and policy-driven water management. The city's water use composition remained stable, with industry being the largest consumer, although its share gradually decreased.
(2) SD predictions reveal diverging trends in future water demand under different development modes. Under scenarios of accelerated economic growth, water demand increases significantly, highlighting the impact of economic activity on water resources. In contrast, historical and population growth scenarios show a slow decline in demand.
(3) In 2022, the Yangtze River's Datong station showed flood-season drought characteristics, with conditions more severe than those of the extreme droughts of 1978 and 2006. This exacerbated saltwater intrusion and increased the number of days that reservoirs were unusable, revealing vulnerabilities in Shanghai's water supply system.
(4) WSS risks decrease as the initial reservoir water level rises, but water shortages persist in drought conditions. Risks will gradually decrease in the business-as-usual mode but increase with economic acceleration and fluctuate in population acceleration and comprehensive development modes. By 2050, the scenarios

with the lowest and highest WSS risks project WSIs, DSA, and WSA of 1.33–2.07, 33–88 days, and 343–896 million m^3, respectively.

(5) Future strategies should focus on improving water use efficiency to manage demand levels. Enhancing smart monitoring and upstream reservoir regulation is crucial. Short- and mid-term strategies suggest prioritizing local and transit water resources and activating emergency water sources based on efficiency and reasonable reservoir water levels. Long-term approaches recommend new reservoir constructions and inter-regional water transfers to meet extreme-event demands.

This study presents an innovative framework for assessing WSS risks in Shanghai and similar estuarine cities, using a combination of SD and DAPP. This approach allows for a comprehensive assessment of the interplay between socio-economic factors, water use, and climate change impacts, enhancing strategic planning for water security. This study provides actionable strategies for water resource management in Shanghai, which could guide policy adjustments and infrastructure development to enhance the city's resilience against future water scarcity. It also offers valuable insights for other large estuarine cities worldwide facing similar challenges due to extreme climatic conditions. However, the primary limitation is the reliance on the 2022 extreme drought event for risk assessment, without considering a broader range of future climate scenarios. Additionally, we did not consider the impact of water scarcity on socio-economic factors when building the SD model. This study focuses on the highest-risk scenario for the DAPP map without quantifying its cost-effectiveness. Future work will aim to expand the analysis to include different climate conditions, such as sea level rise and storm surges, as well as water scarcity constraints and their impacts on WSS, to provide a more detailed perspective on managing water resources in a changing climate.

Chapter 9

Suggestions for Future Research

9.1 Recommendation to Expedite the Research and Development and the Construction of Smart, High-Quality Water Network Infrastructure in the Yangtze River Delta to Face the Challenges Posed by Extreme Climate Change and Emergencies

The urban agglomeration in the Yangtze River Delta (YRD) region occupies an important position in China's and even the world's economic landscape. Bordering the Yellow Sea and the East China Sea, it covers 41 cities in Shanghai, Jiangsu, Zhejiang, and Anhui, with a total population of 237 million and a total area of 358,000 km^2. With less than 4% of the country's land area, it has consistently maintained about a quarter of the country's economic output. However, the region's water resources support for high-quality economic and social development is inadequate. The per capita water resource volume is only 940 m^3, far below the world average. The quality of raw water for urban and rural water supplies is poor, and the water supply, flood control, and coastal defense capabilities are weak.

In recent years, global warming has become increasingly evident. The global average temperature has increased by an average of 0.15°F per decade since 1901, reaching the highest temperature on meteorological record in 2023, and is projected to rise by 1.5–2.8°C by the end of this century. The surface temperature of the world's oceans has gradually increased since the 20th century, with an increase of 0.13°F per decade from 1901 to 2015. The global average sea level has risen at a rate of 1.52 mm per year from 1880 to 2013, and since 2014, this rate has increased to 2.79–3.81 mm per year, almost double the previous rate. Meanwhile, China has experienced more than 50 high-impact rainfall and flood events and over 20 high-impact typhoons, droughts, and extreme heatwaves since the 21st century. This series of observational evidence highlights that China's weather and climate conditions are developing toward a new normal of frequent and intense extreme events, with some historically low-probability extreme events becoming more normalized.

In 2020, the Yangtze River basin experienced a basin-wide flood, the second most severe since the founding of the People's Republic of China (1949), following those in 1954 and 1998. The Taihu Lake basin, which directly affects Shanghai, has experienced floods in 1991, 1999, and 2020 over the past 30 years. Major droughts in the Yangtze River basin in 1961, 1978, and 2022 mean that we cannot rule out the possibility of extreme and persistent drought events like those in 1627–1643, 1874–1879, and 1922–1932 occurring in the future. During the rare drought in the Yangtze River basin in 2022, Shanghai's Qingcaosha Reservoir experienced 98 consecutive days of undrinkable water due to extreme saltwater intrusion, far exceeding the design standard of 68 days. Chenxing Reservoir experienced 11 saltwater intrusions between September 2006 and March 2007. During Typhoon In-Fa in 2021, Shanghai's Bund was severely flooded. Some areas in the YRD region have poor water network connectivity. Floods and droughts in the Taihu Lake basin and the Huai River basin have historically posed significant threats, placing considerable pressure on the water environment and ecological protection and resulting in inadequate shipping capacity. These issues pose serious threats to regional water resource security, including flood and drought losses, water supply crises, water environment and ecology damage, and disruptions to inland river shipping, challenging the sustainable socio-economic development of the region. Therefore, there is an urgent need for regional water network construction, and the region possesses favorable conditions and obvious feasibility advantages.

On the other hand, the frequent occurrence of droughts and floods in the YRD in recent years indicates that the existing flood and drought control and water supply systems are struggling to cope with extreme climate events. The current forecasting level cannot achieve ultra-long-term forecasts, and modern urban life and industries cannot withstand extreme droughts and floods. Therefore, we need to adopt innovative thinking to improve the adaptability of the water resource system in the YRD for managing extreme climate events. The smart water network is a digital water network that integrates spatial data, process data, and topological data. It can be constructed and iteratively evolved with the support of new-generation information technologies, such as big data, cloud computing, the Internet of Things, and artificial intelligence, to provide water networks in various forms to meet the application needs of the business layer, ultimately achieving the goals of real-time monitoring, rapid transmission, accurate forecasting, optimal decision-making, precise allocation, and efficient management. It can provide support for precise allocation, optimal decision-making, and efficient management of water resources during droughts and floods. It is recommended to accelerate the construction of a smart water network with fully independent intellectual property rights in the YRD region to ensure sustainable water resource engineering strategies that address urgent needs within world-class strategic emerging industrial clusters in the YRD and promote the YRD regional integration strategy.

Following thorough investigation and summarization of flood and tide control and water supply situations in major international estuarine cities such as New York, Greater Vancouver, London, Greater Paris, Greater Sydney, and Tokyo, as well as the roles and issues of existing water conservancy projects like the Foziling Reservoir in Huaibei and the Qingcaosha Reservoir in Shanghai, the Huangpu River-Taipu River regulating sluice in Shanghai, the north and south diversion rivers on Chongming Island, the sluice and dams along the Shaying River in Anhui's Huaibei, water diversion projects from the Yangtze River to the Huai River and Taihu Lake, and the Pisihang Irrigation District, and considering the demand for flood and drought control in water network construction, it is recommended to expedite research and construction of a risk over-limit perception and analysis system for the infrastructure of the smart and high-quality water network in the YRD. This specifically includes the following three important aspects.

9.1.1 *Research and construction of China's space reference antenna*

In the context of the world's major river delta regions, accurately quantifying the magnitude and impact of sea-level rise poses a global challenge in addressing climate change and sea-level rise. Based on the research findings and recommendations of academicians (experts), the Shanghai Municipal Government invested heavily in 1996 and 2013 to implement regional and local engineering and non-engineering measures to combat climate change, achieving strategic defensive outcomes. In recent years, numerous research achievements regarding sea-level changes have been made, utilizing global, regional, and local high-temporal and spatio-resolution information on sea-level and its variations provided by tide gauge stations, satellites, buoys, submersibles, radars, and other sensing devices. However, the lack of unified of temporal and spatial benchmarks for monitoring sea-level changes at these scales has hindered their practical application in regional and local engineering responses.

Currently, the establishment of temporal and spatial benchmarks internationally relies primarily on ground-based observations. The correction capabilities for various terrestrial gravitational and deformation effects have essentially reached their limits, making it difficult to further improve precision. Consequently, the current highest precision for temporal and spatial benchmarks is approximately nanoseconds (for time) and centimeters (for space). Multi-technology colocation and the space-based transformation of temporal and spatial benchmarks will fundamentally overcome the constraints of existing observations, thereby achieving indicator enhancements and expanding scientific applications.

9.1.2 *Research and development of integrated sky–ground–sea monitoring data transmission technology and equipment*

The YRD watershed covers an area of 23,937.2 km^2, with mainland coastlines of 213 km, 954 km, and 2,253 km in Shanghai, Jiangsu, and Zhejiang, respectively. Currently, the monitoring and sensing of the water system and offshore areas primarily rely on 4G/5G terrestrial communication networks, but several issues remain:

(1) While full 4G/5G coverage has been achieved in the Jiangsu section of the Yangtze River and the Taihu Lake section, other areas are still under limited coverage, resulting in numerous blind spots. This makes it difficult to conduct large-scale, multi-frequency, and high-precision monitoring of hydrology, water quality, algae, and other factors, as well as providing early warnings for emergencies.

(2) Extreme weather events such as floods and typhoons can cause water ingress, power outages, and fiber optic cable disruptions at base stations, leading to service disruptions in 4G/5G networks. For example, during heavy rainfall in Changzhou in 2016, over 100 base stations were flooded and lost power, exiting service. In 2021, Henan Province experienced a cumulative total of 61,900 base stations ceasing operation due to rainstorm disasters. In 2023, the Beijing–Tianjin–Hebei region experienced heavy rainfall, with Beijing Mobile alone reporting 1,600 base stations not operational.

(3) Due to blind spots in wireless communication network coverage or low communication rates of high-orbit satellites, offshore monitoring and maintenance are still conducted manually, resulting in poor real-time performance. Therefore, it is recommended to expedite research on the construction scheme for an integrated sky–ground–sea intelligent monitoring network in the YRD and adjacent sea areas and develop prototype terminals for low-orbit satellites with integrated communication, navigation, and remote sensing capabilities, achieving a peak star-to-ground rate exceeding 150 Mbps. Furthermore, an on-board implementation plan for multi-dimensional sensing data tailored to the YRD and adjacent sea areas should be proposed, and transmission standards for sensing data should be formulated in these regions. These efforts will revolutionize monitoring from multi-spectral to hyperspectral and from scattered points to large areas, enabling high-precision estimation of 10 water quality indicators—water temperature, pH, total nitrogen, ammonia nitrogen, total phosphorus, chlorophyll-a, CODCr, CODMn, dissolved oxygen, and total suspended solids—and early warning for events such as

red tide outbreaks. This will significantly promote cost reduction and efficiency enhancement in water monitoring within the YRD, thereby supporting the protection and high-quality development of the YRD water system.

9.1.3 *Research and development of the AIoT system for the Yangtze River Delta water network and a large-scale model for predicting the spatio-temporal distribution of water resources*

The YRD region features a low and flat terrain, where factors such as transit floods, regional heavy rainfall, storm surges, and saltwater intrusion overlap, leading to flood disasters and difficulties in extracting water from groundwater sources. Meanwhile, the YRD boasts a developed economy and a high population and industrial concentration, with key estuaries such as the Yangtze River Estuary serving as primary outlets for pollution within the basin. Although China has kept pace with international standards in traditional water resource ecological monitoring and scheduling, with an increasing variety of equipment and methods yielding remarkable results and essentially possessing the foundational layer of IoT sensing and information transmission, IT infrastructure, and a closed-loop system of "sensing-transmission-perception-application," there are still significant issues. Specifically, the informatization systems of water network projects are fragmented and decentralized, lacking an integrated operational and management model. Data isolation is severe, with platform-based intensive construction being limited, scattered, and inadequate, resulting in numerous shortcomings in comprehensive sensing, precise control, and intelligent decision-making. Predictions, simulations, and projection based on digitization and artificial intelligence have not yet been fully realized. The capabilities of unified, precise, and multi-level linked scheduling do not match or adapt to the construction of the YRD water network in the new era, nor to achieving cross-regional collaborative scheduling and efficient operation.

The aforementioned issues involve the core challenges of the current smart water network, namely algorithms and models consisting of basic data sets and model sets. In terms of basic data sets, there is an urgent need to comprehensively utilize domestic and international satellite observations, socio-economic and statistical data, and Earth big data from various sources to conduct case studies on smart and high-quality water networks in the YRD at regional, provincial, municipal, and county scales. The goal is to develop basic data sets with independent intellectual property rights that can be publicly shared and meet the needs of indicator monitoring and assessment, while improving construction standards for key artificial estuary hydrological stations below the Datong Hydrological Station. Among them, it is particularly important to improve the construction standards of the Xuliujing Hydrological Control Station, as this directly ensures the accuracy of basic hydrological and sediment data for flood and tide prevention, water supply, port and shipping, airports, and other critical infrastructure in Shanghai, a leading city in the Yangtze River Economic Belt. The urgency of this matter cannot be overstated.

Regarding model sets, most studies currently use foreign numerical models, and there is an urgent need to develop a large-scale model with independent intellectual property rights for predicting the spatio-temporal distribution of water resources. The input for this model includes online monitoring based on AIoT and collaborative, multi-source, heterogeneous remote sensing big data from space, air, land, and sea, online sensor monitoring data, high-precision spatio-temporal data products, spatio-temporal sequence monitoring data, related documents and texts, multi-source remote sensing images, and surveillance videos of the YRD water network at both macro and local scales. This large-scale model constructs a multi-modal structure to deeply understand the impact of information hidden in text and images on spatio-temporal distribution data of water resources. It is equipped with a real-time scene data invocation mode to analyze and predict the spatio-temporal distribution of water resources. Meanwhile, the model utilizes a large-model thinking chain to invoke multiple agents, including water environment and ecological disaster models such as algal blooms and saltwater intrusion, drawing on a deep-learning-based offline water resource scheduling rule base to provide decision-makers with more comprehensive countermeasures and solutions in various scenarios. This enables online assessments and early warnings of security risks in the high-quality water network of the YRD and simulates the actual effects of policy implementation. Finally, resource scheduling is conducted through the identification of trigger events such as population density and disaster response. Overall, the large-scale model for predicting the spatio-temporal distribution of water resources serves as the infrastructure, complemented by scenario-specific applications such as policy implementation simulations.

9.2 Establish a Comprehensive Atlas for Water Networks Early to Address New Challenges in National Water Network Construction and Strengthen the Micro-Storage and Micro-Circulation Capabilities of Shanghai's Modern Water Network

Modern water networks are integrated water distribution systems based on natural water systems, with water diversion and drainage projects serving as channels and control structures at nodes as hubs. They combine functions such as flood and drought control, water resource allocation, water ecological protection, and water transportation. Composed of natural rivers, lakes, and multiple engineering projects, modern water networks are not only water resource networks but also water ecological, water energy, and water transportation networks. They are complex network systems with combined functions, difficult scheduling, and significant spatial and temporal scale effects. According to the "Outline of the National Water Network Construction Plan," modern water networks can be divided into four levels based on administrative divisions: national backbone networks, provincial water networks, municipal water networks, and county-level water networks. Functionally, modern water network projects can be divided into two levels: national backbone water network projects and regional water network projects.

In the "Shanghai Water Network Construction Plan," approved by the Shanghai Municipal People's Government in February 2024, it was proposed to accelerate the connection of Shanghai's water network with the national water network to ensure water resource security, improve flood control and disaster reduction capabilities, optimize the water ecological environment, and promote the development of smart water services. This will provide robust water security for Shanghai, accelerating its development into a socialist, modern, international metropolis with global influence. The "Shanghai Water Network Construction Plan" outlines an overall layout of "Three Rivers, One Network, and Ten Hubs," following the principle of "receiving upstream water from river basins, connecting downstream to rivers and seas, following nature, and adapting to development." The primary tasks of current water network construction in Shanghai are divided into four key areas: constructing a safe and high-quality water source allocation system with interconnection and mutual support; improving flood defense and disaster reduction systems for water safety and resilience; creating a network of healthy rivers and lakes that prioritize ecology and harmonize human–water relations; and building a shared, efficient, and smart digital twin water network.

Shanghai's water network is crucial infrastructure for comprehensively addressing issues related to water resources, water ecology, water environment, and water security and for comprehensively enhancing Shanghai's water security capability. To build Shanghai's water network, it is necessary to proceed from a broad perspective that aligns with major national and Shanghai development strategies, as well as the development, governance, and protection of the Yangtze River Economic Belt and the YRD. Long-term and scientific planning should be adopted, and the new generation of information technology should be fully utilized to digitally empower all aspects of the planning, design, construction, and operation and maintenance of Shanghai's water network. This will enhance the digitization, networking, and intelligence of Shanghai's water network, ultimately providing robust water security support for comprehensively building a modern socialist country.

Due to the limitations of Shanghai's natural resource endowment, the current construction of modern water networks in the city faces the following 11 major challenges: (1) It is affected by the operation of the Three Gorges Reservoir upstream and by the ocean, especially sea-level rise, downstream; (2) although the total regional precipitation is abundant, balancing water resources is difficult; (3) the quality of urban and rural water supply is poor, and the water supply's anti-risk capability is weak; (4) rapid socio-economic development high demands for water environment and ecological protection; (5) the plum rain season is prolonged, and the impact of typhoons is significant; (6) he connectivity of water networks is insufficient in some areas; (7) lakes and reservoirs, key components of the water network and vital for regulation and storage, are inadequate, hindering the effectiveness of capillary water networks' micro-storage and micro-circulation functions; (8) historically developed inland water transportation within the region needs to be restored; (9) the water network in the plain area is complex, with reciprocal flow, and ground subsidence significantly impacts the network; (10) flood drainage capacity is insufficient, with singular and inconsistent pathways; and (11) drought relief measures are inadequate, and monitoring capabilities and means need to be improved since they fall short of smart requirements. To address these challenges, the following suggestions are provided.

9.2.1 *Strengthening the investigating Shanghai's water network and constructing a complete atlas of Shanghai's water network*

First, enhance the investigation of the water network. At the municipal level in Shanghai, it is necessary to conduct thorough investigations, analyses, and research on the city's water network and its adjacent provinces, clarifying the intricate internal relationships within the water network system of the estuarine delta metropolis (Shanghai), which is characterized by complex controlling factors, a vast hierarchical system, and intricate correlations.

Secondly, construct a complete lineage of Shanghai's water network. Based on the investigation of the water network in Shanghai and its neighboring areas, a comprehensive atlas of Shanghai's water network will be developed. This will involve identifying and overcoming key bottlenecks that restrict the micro-storage and micro-circulation functions of Shanghai's water network, ultimately providing support for solving significant theoretical and engineering practice issues in the development of the network (including multi-dimensional balanced regulation of water resources in megacities; water resource allocation methods considering multi-objective feedback and multi-agent decision-making preferences; real-time dynamic regulation methods for water resources considering forecast uncertainty; and coordination and win-win mechanisms for user demand conflicts under extreme conditions.

9.2.2 *Exploring the high-precision digital water network timeline and evolutionary patterns of Shanghai over the past 30 years*

This involves exploring and analyzing the sequential characteristics of Shanghai's water network lineage's evolution over time and its co-evolutionary relationships with neighboring provinces. Efforts will be intensified to develop a high-precision timeline representation model for digital water networks, incorporating isochronal comparison significance, as well as natural environment, administrative divisions, and socio-economic factors. Precise artificial intelligence methods are needed for social development—specifically, water network prediction—based on deep learning, water network atlases, and reasoning engines to deduce the evolutionary laws and development directions of Shanghai's water network. Ultimately, a representation framework and a general pattern for water network atlases that integrate temporal and spatial characteristics and complex water network structures will be formed, enabling the intrinsic logical associations and structured expressions of multi-level water networks (including national backbone networks, provincial water networks, municipal water networks, county-level water networks, and village-level water networks). within a unified temporal and spatial dimension.

9.2.3 *Promoting modern water network construction planning guided by the dynamic atlas of Shanghai's water network*

Based on the current status and future development needs of Shanghai's water network, research and implementation will be conducted on the integration, fusion, and correlation technologies between village-level capillary water networks—with micro-storage and micro-circulation as their primary functions—with national-level arterial water networks. Research and optimization must be carried out on diversified water network configurations, including local and transferred water, conventional and unconventional water sources, and routine and emergency water supply systems in Shanghai, promoting the gradual formation of a water supply guarantee system that emphasizes large, medium, small, and micro water networks and complements water sources. Ultimately, this will comprehensively enhance the three major safety guarantee capabilities of water resources, water supply, and water ecology.

9.3 Developing International Joint Research

As global climate, oceanic, and environmental changes accelerate, along with the impact of human activities, challenges and risks related to water resources, water environments, floods, droughts, and water security are becoming increasingly severe. To fully leverage the role of space–air–ground observation and geographic information science and technology, improve water conservancy infrastructure and modernize the water conservancy network system, and enhance the digitization, networking, and intelligence levels of water governance and management, it is urgent to empower smart water conservancy with geographic information technology. This requires cooperation among leaders, experts, and representatives from relevant international organizations, industry authorities, scientific research institutions, universities, and enterprises across Asia, Europe, North America, South America, Australia, and Oceania, as well as China, to jointly participate and address water disaster issues from new perspectives, paradigms, technologies, and methodologies.

Bibliography

Adediran AO, Parcharidis I, Poscolieri M, & Pavlopoulos K. Computer-assisted discrimination of morphological units on north-central Crete (Greece) by applying multivariate statistics to local relief gradients. *Geomorphology*, 2004, 58(1): 357–370.

Aerts JCJH, Botzen WJW, Emanuel K, Lin N, Moel HD, & Michel-Kerjan EO. Evaluating flood resilience strategies for coastal megacities. *Science*, 2014, 344(6183): 473–475.

Ahmad I. Digital elevation model (DEM) coupled with geographic information system (GIS): An approach towards erosion modeling of Gumara watershed, Ethiopia. *Environmental Monitoring and Assessment*, 2018, 190: 1–11.

Ahmadi MS, Susnik J, Veerbeek W, & Zevenbergen C. Towards a global day zero? Assessment of current and future water supply and demand in 12 rapidly developing megacities. *Sustainable Cities and Society*, 2020, 61, 102295.

Ahmed AN, Othman FB, Afan HA, Ibrahim RK, Fai CM, Hossain MS, & Elshafie A. Machine learning methods for better water quality prediction. *Journal of Hydrology*, 2019, 578, 124084.

Alebregtse NC, & de Swart HE. Effect of a secondary channel on the linear tidal dynamics in a semi-enclosed channel: A simple model. *Ocean Dynamics*, 2014, 64: 573–585.

Aliotta S, & Perillo GME. A sand wave field in the entrance to Bahia Blanca estuary, Argentina. *Marine Geology*, 1987, 76: 1–14.

Alizadeh MJ, Kavianpour MR, Kisi O, & Nourani V. A new approach for simulating and forecasting the rainfall-runoff process within the next two months. *Journal of Hydrology*, 2017, 548: 588–597.

Allen JRL. *Sedimentary Structures: Their Character and Physical Basis, Developments in Sedimentology 30*. Elsevier Science Publishers, 1982: Amsterdam 593.

Allen GH, & Pavelsky TM. Global extent of rivers and streams. *Science*, 2018, 361: 585–588.

Allison MA, Demas CR, Ebersole BA, Kleiss CA, Little CD, Meselhe EA, Powell NJ, Pratt TC, & Vosburg BM. A water and sediment budget for the lower Mississipi-Atchafalaya River in flood years 2008–2010: Implications for sediment discharge to the oceans and coastal restoration in Louisiana. *Journal of Hydrology*, 2012, 432–433(8): 84–97.

Almeida LP, Almar R, Bergsma EW, Berthier E, Baptista P, Garel E, Dada OA, & Alves B. Deriving high spatial-resolution coastal topography from sub-meter satellite stereo imagery. *Remote Sensing*, 2019, 11(5), 590.

Arns A, Wahl T, Dangendorf S, & Jensen J. The impact of sea level rise on storm surge water levels in the northern part of the German Bight. *Coastal Engineering*, 2015, 96(2):118–131.

Ashley GM. Classification of large-scale subaqueous bed form: A new look at an old problem. *Journal of Sedimentary Petrology*, 1990, 60(1): 160–172.

Azma A, Liu Y, Azma M, Saadat M, Zhang D, Cho J, & Rezania S. Hybrid machine learning models for prediction of daily dissolved oxygen. *Journal of Water Process Engineering*, 2023, 54, 103957.

Bação F, Lobo V, & Painho M. The self-organizing map, the Geo-SOM, and relevant variants for geosciences. *Computers & Geosciences*, 2005, 31(2): 155–163.

Bakker K. Water security: Research challenges and opportunities. *Science*, 2012, 337: 914–915.

Bamber JL, Oppenheimer M, Kopp RE, Aspinall WP, & Cooke RM. Ice sheet contributions to future sea-level rise from structured expert judgment. *Proceedings of the National Academy of Sciences*, 2019, 116(23): 11195–11200.

Barbier EB, Hacker SD, Kennedy C, Koch EW, Stier AC, & Silliman BR. The value of estuarine and coastal ecosystem services. *Ecological Monographs*, 2011, 81(2): 169–193.

Barnard PL, Erikson LH, & Kvitek RG. Small-scale sediment transport patterns and bedform morphodynamics: New insights from high-resolution multibeam bathymetry. *Geo-Marine Letters*, 2011, 31: 227–236.

Baruch A, Filin S. Detection of gullies in roughly textured terrain using airborne laser scanning data. *ISPRS Journal of Photogrammetry and Remote Sensing*, 2011, 66(5): 564–578.

Belgiu M, & Draguţ LD. Random forest in remote sensing: A review of applications and future directions. *ISPRS Journal of Photogrammetry and Remote Sensing*, 2016, 114: 24–31.

Bernardara P, Mazas F, Kergadallan X, & Hamm L. A two-step framework for over threshold modelling of environmental extremes. *Natural Hazards Earth System Science*, 2014, 14: 635–647.

Bertels L, Houthuys R, Sterckx S, Knaeps E, & Deronde B. Large-scale mapping of the riverbanks, mud flats and salt marshes of the Scheldt basin, using airborne imaging spectroscopy and LiDAR. *International Journal of Remote Sensing*, 2011, 32(10): 2905–2918.

Best J. The fluid dynamics of river dunes: A review and some future research directions. *Journal of Sedimentary Petrology*, 1990, 60(1): 160–172.

Blondeaux P, & Vittori G. A model to predict the migration of sand waves in shallow tidal seas. *Continental Shelf Research*, 2016, 112: 31–45.

Bolongaro-Crevenna A, Torres-Rodríguez V, Sorani V, Frame D, & Ortiz M A. Geomorphometric analysis for characterizing landforms in Morelos State, Mexico. *Geomorphology*, 2005, 67(3–4): 407–422.

Bonì R, Bordoni M, Colombo A, Lanteri L, & Meisina C. Landslide state of activity maps by combining multi-temporal A-DInSAR (LAMBDA). *Remote Sensing of Environment*, 2018, 217: 172–190.

Breiman L. Random forests. *Machine Learning*, 2001, 45, 5–32.

Bridge JS, & Jarvis J. Flow and sedimentary processes in the meandering river South Esk, Glen Clova, Scotland. *Earth Surface Processes & Landforms*, 2010, 1(4): 303–336.

Brunt, KM, Neumann, TA, & Smith, BE. Assessment of ICESat-2 ice sheet surface heights, based on comparisons over the interior of the Antarctic ice sheet. *Geophysical Research Letters*, 2019, 46, 13072–13078.

Buchanan MK, Oppenheimer M, & Kopp RE. Amplification of flood frequencies with local sea level rise and emerging flood regimes. *Environmental Research Letters*, 2017, 12(6), 064009.

Burchard H, & Hetland RD. Quantifying the contributions of tidal straining and gravitational circulation to residual circulation in periodically stratified tidal estuaries. *Journal of Physical Oceanography*, 2010, 40:1243–1262.

Burrough PA, van Gaans PFM, & MacMillan RA. High-resolution landform classification using fuzzy k-means. *Fuzzy Sets and Systems*, 2000, 113(1): 37–52.

Cai R, Li B, Fang W, Fu S, Han Z, Han J, & Tan H. On the comprehensive risk of global change in coastal zone and coastland of China. *China Basic Science*, 2017, 6: 24–29.

Cai Y, & Wen Y. Influence of sea level rise on the determination of top surface elevation of coastal projects. *Water Transportation Engineering*, 2015(7): 87–91.

Cao L, XU J, Li G, & Shi J. High-resolution morphological characteristics of sand waves off the west Hainan Island. *Marine Geology and Quaternary Geology*, 2006, 26(4): 15–22.

Cataño-Lopera YA, Abad JD, & García MH. Characterization of bedform morphology generated under combined flows and currents using wavelet analysis. *Ocean Engineering*, 2009, 36(9–10): 617–632.

Cazenave PW, Dix JK, Lambkin DO, & McNeill LC. A method for semi-automated objective quantification of linear bedforms from multi-scale digital elevation models. *Earth Surface Processes and Landforms*, 2013, 38: 221–236.

Cha SH. Comprehensive survey on distance/similarity measures between probability density functions. *International Journal of Mathematical Models & Methods in Applied Sciences*, 2007, 1(4): 300–307.

Chea H, & Sharma M. Residential segregation in hillside areas of Seoul, South Korea: A novel approach of geomorphons classification. *Applied Geography*, 2019, 108: 9–21.

Chen C, Zhang C, Tian B, Wu W & Zhou Y. Tide2Topo: A new method for mapping intertidal topography accurately in complex estuaries and bays with time-series Sentinel-2 images. *ISPRS Journal of Photogrammetry and Remote Sensing*, 2023, 200, 55–72.

Chen F, Fu Z, & Yang F. Impact of river regime changes in the Zhenjiang reach of the Yangtze River on navigation conditions. *Water Transport*, 2011, (6): 112–116.

Chen J. Topographic development of the Jiangkou section of the Yangtze River Delta. *Acta Geographica Sinica*, 1957, 24(3): 241–253.

Chen J. Developing shallow beach and tideland resources to expand China's living space. *Chinese Engineering Science*, 2000, 2(3): 27–31.

Chen J, & Chen S. Challenges faced by Chinese estuaries and coasts. *Frontiers of Marine Science*, 2002, 18(1): 1–5.

Chen J, Cheng H, & Dai Z. The role and response of the third driving force in estuarine processes: A case study of the Yangtze River Estuary. *Advances in Natural Science*, 2008, 18(9): 994–1000.

Chen J, Liu X, Zheng Z, & Song W. Enhancing effect of ground subsidence on storm surge disasters. *Shanghai Land and Resources*, 2014(4): 134–136.

Chen J, Wang Y, & Cai H. Study on the profile characteristics of tidal flats along the coast of Jiangsu. *Ocean Engineering*, 2010, (4): 94–100.

Chen J, & Yun C. Evolution of the Yangtze River channel between Nanjing and Wusong. *Acta Geographica Sinica*, 1959, (3): 55–73.

Chen J, Yun C, Xu H, & Dong Y. Patterns of Yangtze River Estuary development over two millennia. *Acta Oceanologica Sinica*, 1979, 1(1): 103–111.

Chen M. Discussion on ecological river management in Chongming District. *Shanghai Water Affairs*, 2018, (1): 45–47.

Chen W, Jin Z, Yu Z, *et al.* Prediction of COD(Mn) in Poyang Lake based on empirical wavelet transform. *Journal of Environmental Engineering Technology*, 2023, 13(01): 180–187.

Chen W, Jin Z, Yu Z, *et al.* Prediction of COD in Poyang Lake based on a two-layer data decomposition hybrid model. *Transactions of the Chinese Society of Agricultural Engineering*, 2022, 38(05): 296–302.

Chen W, Li X, Wang Y, Chen G, & Liu S. Forested landslide detection using LiDAR data and the random forest algorithm: A case study of the Three Gorges, China. *Remote Sensing of Environment*, 2014, 152: 291–301.

Chen W, & Prior D. Discussion on the microgeomorphology and its causes of the submarine slope in the Yellow River Estuary. *Journal of Ocean University of Qingdao: Natural Science Edition*, 1992, 22(1): 71–81.

Chen W, Van Gent MRA, Warmink JJ, & Hulscher SJMH. The influence of a berm and roughness on the wave overtopping at dikes. *Coastal Engineering*, 2020, 156:1–29.

Chen W, Yang Z, Cao L, Guo Z, Liu H, & Du J. Types and zoning of microgeomorphology on the submarine slope of the modern Yangtze River Estuary. *Journal of Ocean University of China: Natural Science Edition*, 1993, (S1): 45–51.

Chen Y. Study on the spatiotemporal variation of water environment quality in the Jiangsu section of the Beijing-Hangzhou Grand Canal. Nanjing Normal University, 2005.

Chen Z, Li J, & Tan X. Study on the development strategy of urban water resources in Shanghai. *China Water & Wastewater*, 2018, 34(02): 24–30.

Chen Z, Ruan R, Han C, Shen H, Chen G, & Xia X. Emergency response plan system for saline tide invasion in the Yangtze River estuary water source area. *Water Resources Protection*, 2019, 35(5): 52–58.

Chen Z, Shi X, & Mao X. Research on the impact of sea level rise on drainage safety of Chongming Islands in the Yangtze River Estuary. *Climate Change Research Progress*, 2015, 11(4): 239–244.

Chen ZL, & Lee SY. Tidal flats as a significant carbon reservoir in global coastal ecosystems. *Frontiers in Marine Science*, 2022, 9, 900896.

Cheng H, Chen J, Chen Z, Ruan R, Xu G, Zeng G, Zhu J, Dai Z, & Chen X. Research on the impact of sea level rise on the Yangtze River Estuary. *China Science and Technology Achievements*, 2017, 18(3): 27–29, 34.

Cheng H, Hu H, Jiang Z, & Wang B. Spectral analysis of bottom form balance in the eastern passage of the Qiongzhou Strait. *Ocean Engineering*, 2003, 21(4): 97–103.

Cheng H, & Jiang Y. *Physical Processes in the Middle and Lower Reaches of the Yangtze River*. Beijing: Science Press, 2021. p. 712.

Cheng H, Li M, Zhou T, & Xue Y. High-resolution submarine microgeomorphology and movement characteristics of the Yangtze River Estuary. *Ocean Engineering*, 2002, 20(2): 91–95.

Cheng H, Shi Z, Kastasckuk R, & Dong L. Stability domain of sand waves in the South Branch of the Yangtze River Estuary. *Oceanology and Limnology*, 2004, 35(3): 214–220.

Cheng H, Song B, Xue Y, & Mao X. Study on the transport characteristics of coarse silt and ultrafine sand in the Yangtze river estuary—episodic resuspension and bed movement. *Marine Geology and Quaternary Geology*, 2000, (1): 20–27.

Cheng H, Tana Z, Zhou Y, Zhu J, Ruan R, & Chen J. Risk assessment and countermeasures for water supply safety of Shanghai's Yangtze river estuary water source under the background of sea level rise. *Climate Change Research Progress*, 2015, 11(4): 263–269.

Cheng HQ, & Chen JY. Adapting cities to sea level rise: A perspective from Chinese deltas. *Advances in Climate Change Research*, 2017, 8(2): 130–136.

Cheng HQ, Chen JY, Chen ZJ, Ruan RL, Xu GQ, Zeng G, Zhu JR, Dai ZJ, Chen XY, Gu SH, Zhang XL, & Wang HM. Mapping sea level rise behavior in an estuarine delta system: A case study along the Shanghai coast. *Engineering*, 2018, 4 (1): 156–163.

Cheng HQ, Kostaschuk R, & Shi Z. Tidal currents, bed sediments, and bedforms at the South Branch and the South Channel of the Changjiang (Yangtze) estuary, China: Implications for the ripple-dune transition. *Estuaries*, 2004, 27(5): 861–866.

Cheng HQ, Li JF, Yin DW, Li MT, & Wang B. Nearshore bedform instability in the eastern entrance to the Qiongzhou Strait, South China Sea. *Frontiers of Earth Science in China*, 2008, 2(3): 283–291.

Cheng HQ, Teng L, & Chen W. Dune dynamics in coarse silt, sand and gravel along the main channel from the estuarine front of the Yangtze River to the Three Gorges Dam, April 1–3, 2019. *Proceedings of International Conference Maine and River Dune Dynamics (MARID VI)*, 2019.

Cheng J, & Shen B. Remote sensing study of bank collapse in the Wanjiang section of the Yangtze River. *Journal of Anhui Normal University (Natural Science Edition)*, 1995, 18(2): 84–88.

Clark PU, Mix AC, Eby M, Levermann A, Rogelj J, Nauels A, & Wrathall DJ. Sea-level commitment as a gauge for climate policy. *Nature Climate Change*, 2018, 8: 653–655.

Colesanti C, & Wasowski J. Investigating landslides with space-borne Synthetic Aperture Radar (SAR) interferometry. *Engineering Geology*, 2006, 88(3–4): 173–199.

Colosimo I, Sebastiaan D, & Lodewijk P, Johan Christiaan Winterwerp and Prooijen, van. Winds of opportunity: The effects of wind on intertidal flat accretion. *Geomorphology*, 2023, 439, 108840–108840.

Cooper HM, Fletcher CH, Chen Q, & Barbee MM. Sea-level rise vulnerability mapping for adaptation decisions using LiDAR DEMs. *Progress in Physical Geography*, 2013, 37, 745–766.

Council N. Informing an effective response to climate change. 2010.

Cui X, Xing Z, Yang F, Fan M, Ma Y, & Sun Y. A method for multi-beam seafloor terrain classification based on self-adaptive geographic classification unit. *Applied Acoustics*, 2020, 157: 107029.

Cui Y, Jiang F, & Ren B. Numerical simulation of internal stress distribution in sloped protective block body under wave action. *Ocean Engineering*, 2014, 32(5): 78–84.

Curcio AC, Peralta G, Aranda M, & Barbero L. Evaluating the performance of high spatial resolution UAV-photogrammetry and UAV-Lidar for salt marshes: The Cádiz Bay study case. *Remote Sensing*, 2022, 14(15), 3582.

da Silva FP, Martins JRS, & Nogueira FF. Impacts of sea level rise on seawater intrusion in Cubato River, Brazil. *Environmental Modeling & Assessment*, 2020, 25(6): 831–841.

Dai W, Li H, Chen X, Xu F, Zhou Z, & Zhang C. Saltmarsh expansion in response to morphodynamic evolution: Field observations in the Jiangsu Coast using UAV. *Journal of Coastal Research*, 2020, 95(SI): 433–437.

Damen J M, Dijk T A G P, Hulscher S J M H. Spatially varying environmental properties controlling observed sand wave morphology. *Journal of Geophysical Research: Earth Surface*, 2018, 123: 262–280.

Dangendorf S, Mudersbach C, Jensen J, Anette G, & Heinrich H. Seasonal to decadal forcing of high water level percentiles in the German Bight throughout the last century. *Ocean Dynamics*, 2013, 63(5): 533–548.

Davison AC, Smith RL. Models for exceedances over high thresholds. *Journal of the Royal Statistical Society: Series B (Methodological)*, 1990, 52(3): 393–442.

Debese N, Jacq JJ, & Garlan T. Extraction of sandy bedforms features through geodesic morphometry. *Geomorphology*, 2016, 268: 82–97.

Deng S, Xia J, Zhou M, & Li J. Simplified simulation of the bank retreat process in the Upper Jingjiang River. *Chinese Science Bulletin*, 2016, 61(33): 3606–3615.

Deng S, Xia J, Zhou M, & Lin F. Coupled modeling of bed deformation and bank erosion in the Jingjiang Reach of the middle Yangtze River. *Journal of Hydrology*, 2019, 568: 221–233.

DeWitt JD, Warner TA, & Conley JF. Comparison of DEMS derived from USGS DLG, SRTM, a statewide photogrammetry program, ASTER GDEM and LiDAR: Implications for change detection. *GIScience & Remote Sensing*, 2015, 52, 179–197.

Dijkstra YM, Brouwer RL, Schuttelaars HM, & Schramkowski GP. The iFlow modelling framework v2.4: A modular idealized process-based model for flow and transport in estuaries. *Geoscientific Model Development*, 2017, 10(7): 2691–2713.

Ding L, Dou X, Gao X, Pan Y, & Jiao Z. A review of saltwater intrusion research in the Yangtze river estuary. Nanning: 20151042–1046.

Dixon MJ, Tawn JA. Extreme sea-levels at the UK A-class sites: Site-by-site analyses. *Proudman Oceanographic Laboratory Internal Document*, 1994, 65.

Dong J, Liao M, Xu Q, Zhang L, Tang M, & Gong J. Detection and displacement characterization of landslides using multi-temporal satellite SAR interferometry: A case study of Danba County in the Dadu River Basin. *Engineering Geology*, 2018, 240: 95–109.

Dong J, Wang Z, Wu J, Cui X, & Pei R. A novel runoff prediction model based on support vector machine and gate recurrent unit with secondary mode decomposition. *Water Resources Management*, 2024, 1–20.

Dong J, Wang Z, Wu J, Huang J, Zhang C. A water quality prediction model based on signal decomposition and ensemble deep learning techniques. *Water Science & Technology*, 2023, 88(10): 2611–2632.

Dou P, Zuo S, Ren Y, Rodriguez MJ, & Dai S. Refined water security assessment for sustainable water management: A case study of 15 key cities in the Yangtze River Delta, China. *Journal of Environmental Management*, 2021, 290: 112588.

Dou X, & Zhu C. Analysis of quality control in the dredging project of Liuxu Kou Channel in the Beijing-Hangzhou Grand Canal. *Transportation Science and Technology*, 2010, (4): 77–79.

Drăguţ L, & Eisank C. Automated object-based classification of topography from SRTM data. *Geomorphology*, 2012, 141–142: 21–33.

Du J, Yang S, & Chen D. Preliminary study on the impact of the Three Gorges project on the evolution of the modern Yangtze River Delta. *Marine Science Bulletin*, 2012, (5): 489–495.

Du X, Li Y, & Gao S. Characteristics of large sand waves, tidal current structure, and suspended sediment transport in the Taiwan Shoal. *Acta Oceanologica Sinica*, 2008, 30(5): 124–136.

Du Y, Zhu Q, Lv Z, Yang Y, & Chen Y. Analysis of the monthly and annual variation characteristics of water discharge into the sea in the Yangtze River Estuary over the years. *Hydropower and Water Resources*, 2019, (10): 8–12.

Duan J, Duan W, & Zhu J. Riverbank collapse and stability analysis. *Journal of Wuhan University (Engineering Edition)*, 2004, 37(6): 17–21.

Duan X, & Xu X. Study on the disaster reduction benefits of coastal flood defense dams against storm surge hazards. *Journal of Disaster Prevention and Reduction*, 2006, 26(3): 279–283.

Duffy G. Patterns of morphometric parameters in a large bedform field: Development and application of a tool for automated bedform morphometry. *Irish Journal of Earth Sciences*, 2012, 30: 31–39.

Eeckhaut MVD, Poesen J, Verstraeten, G, Vanacker V, Moeyersons J, Nyssen J, & van Beek LPH. The effectiveness of hillshade maps and expert knowledge in mapping old deep-seated landslides. *Geomorphology*, 2005, 67: 351–363.

Ehsani AH, Quiel F. Geomorphometric feature analysis using morphometric parameterization and artificial neural networks. *Geomorphology*, 2008, 99(1–4): 1–12.

Ensing E, De Swart HE, & Schuttelaars HM. Sensitivity of tidal motion in well-mixed estuaries to cross-sectional shape, deepening, and sea level rise. *Ocean Dynamics*, 2015, 65(7): 933–950.

Evans IS. General geomorphology, derivatives of altitude and descriptive, 1972. In: Chorley RJ (ed) *Spatial Analysis in Geomorphology*, 1972.

Fan H, Yan H, Teng L, Liu R, Li Z, Cheng H, & Zhang E. The effects of extreme flood events on the turbidity maximum zone in the Yangtze (Changjiang) Estuary, China. *Marine Geology*, 2023, 456.

Fan Z, Ge J, Ding P, & Pan L. Impact of the Yangtze River Estuary Deepwater Channel Project on salinity distribution in the North Channel. *Journal of East China Normal University (Natural Science Edition)*, 2012, 2012(4): 181–189.

Fang C. A two-dimensional mathematical model for water and sediment flow considering the influence of bend circulation. *Journal of China Institute of Water Resources and Hydropower Research*, 2003, 1(3): 190–193.

Fang J, & Shi P. Progress and prospects of coastal flood disaster risk assessment under global climate change. *Progress in Geography*, 2019, 38(5): 625–636.

Fang J, Li Y, Sun Z, & Deng J. Analysis of the variation characteristics of discharge at Datong Station of the Yangtze River. *Hydropower and Energy Science*, 2011, 29(5): 15–18.

Fang JY, Thomas W, Zhang Q, Sanne M, Hu P, Fang J, Du SQ, Dou TF, & Shi PJ. Extreme sea levels along coastal China: Uncertainties and implications. *Stochastic Environmental Research and Risk Assessment*, 2021, 35: 405–418.

Fang Q, Zhao S, & Zhou Y. GIS analysis of the evolution of topography around the Jiuduansha area in the Yangtze River Estuary in the past decade. *Surveying and Spatial Geoinformation*, 2014, (11): 90–93+97.

Farrell SL, Duncan K, Buckley EM, Richter-Menge J, & Li R. Mapping sea ice surface topography in high fidelity with ICESat-2. *Geophysical Research Letters*, 2020, 47.

Fayad I, Baghdadi N, Bailly JS, Barbier N, Gond V, El Hajj M, & Bourgine B. Canopy height estimation in French Guiana with LiDAR ICESat/GLAS data using principal component analysis and random forest regressions. *Remote Sensing*, 2014, 6(12): 11883–11914.

Feng A, Gao J, Wu S, Liu Y, & He X. Research progress on the risk and adaptive strategies of storm surge disasters in China under the background of climate change. *Progress in Geography*, 2016, 35(11): 1411–1419.

Finlayson BL, Barnett J, Wei T, Webber M, Li M, Wang M Y, Chen J, Xu H, & Chen Z. The drivers of risk to water security in Shanghai. *Regional Environmental Change*, 2013, 13(2): 329–340.

Fischer EM, & Knutti R. Observed heavy precipitation increase confirms theory and early models. *Nature Climate Change*, 2016, 6(11): 986–991.

Flörke M, Schneider C, & Mcdonald RI. Water competition between cities and agriculture driven by climate change and urban growth. *Nature Sustainability*, 2018, 1(1): 51–58.

Flynn T, Rozanov A, de Clercq W, Warr B, & Clarke C. Semi-automatic disaggregation of a national resource inventory into a farm-scale soil depth class map. *Geoderma*, 2019, 337: 1136–1145.

Foroumandi E, Nourani V, Huang JJ, & Moradkhani H. Drought monitoring by downscaling GRACE-derived terrestrial water storage anomalies: A deep learning approach. *Journal of Hydrology*, 2023, 616, 128838.

Galvao LS, Arlanche PC, & Dalagnol R. Coupled effects of solar illumination and phenology on vegetation index determination: An analysis over the Amazonian forests using the SuperDove satellite constellation. *GIScience & Remote Sensing*, 2024, 61(1).

Gao B. NDWI—A normalized difference water index for remote sensing of vegetation liquid water from space. *Remote Sensing of Environment*, 1996, 58: 257–266.

Gao W, Shen F, Tan K, Zhang W, Liu Q, Lam NSN, & Ge J. Monitoring terrain elevation of intertidal wetlands by utilising the spatial-temporal fusion of multi-source satellite data: A case study in the Yangtze (Changjiang) Estuary. *Geomorphology*, 2021, 383, 107683.

Gao Z, Han S, Liu K, Zheng Y, & Yu H. Numerical simulation of the impact of mean sea level rise on typhoon storm surges in the East China Sea. *Marine Science Bulletin (English Edition)*, 2008, 10(2): 36–49.

Garlan T. GIS and mapping of moving marine sand dunes, November 15–21 2009. In *Proceedings of the 24th International Cartography Conference*, 2009.

Geertsema M, Blais-Stevens A, Kwoll E, Menounos B, Venditti J G, Grenier A, & Wiebe K. Sensitive clay landslide detection and characterization in and around Lakelse Lake, British Columbia, Canada. *Sedimentary Geology*, 2018, 364: 217–227.

Gesch DB. Best practices for elevation-based assessments of sea-level rise and coastal flooding exposure. *Frontiers in Earth Science*, 2018, 6: 230.

Geyer WR, & Maccready P. The estuarine circulation. *Annual Review of Fluid Mechanics*, 2014, 46: 175–197.

Ghanbari M, Arabi M, Obeysekera J, & Sweet W. A coherent statistical model for coastal flood frequency analysis under nonstationary sea level conditions. *Earth's Future*, 2019, 7(2): 162–177.

Ghorbani Z, Amanipoor H, & Battaleb-Looie, S. Water quality simulation of Dez River in Iran using QUAL2KW model. *Geocarto International*, 2022, 37(4), 1126–1138.

Giddings SN, Fong DA, & Monismith SG. Role of straining and advection in the intratidal evolution of stratification, vertical mixing, and longitudinal dispersion of a shallow, macrotidal, salt wedge estuary. *Journal of Geophysical Research Oceans*, 2011, 116(C3): 3.

Gitelson AA, Gritz YA, & Merzlyak MN. Relationships between leaf chlorophyll content and spectral reflectance and algorithms for non-destructive chlorophyll assessment in higher plant leaves. *Journal of Plant Physiology*, 2003, 160(3): 271–82.

Gou ZL, Wu CP, Song LX, & Wu GY. Discussion on the scour mechanism at the head of a groyne. *Proceedings of the 14th Chinese Ocean (Coastal) Engineering Academic Seminar*, 2009, 1133–1135.

Green A, & Uken R. Submarine landsliding and canyon evolution on the northern KwaZulu-Natal continental shelf, South Africa, SW Indian Ocean. *Marine Geology*, 2008, 254(3–4): 152–170.

Gruber FE, Baruck J, & Geitner C. Algorithms vs. surveyors: A comparison of automated landform delineations and surveyed topographic positions from soil mapping in an Alpine environment. *Geoderma*, 2017, 308: 9–25.

Guerra-Hernández J, Narine LL, Pascual A, González-Ferreiro E, Botequim B, Malambo L, Neuenschwander AL, Popescu SC, & Godinho S. Aboveground biomass mapping by integrating ICESat-2, SENTINEL-1, SENTINEL-2, ALOS2/PALSAR2, and topographic information in Mediterranean forests. *GIScience & Remote Sensing*, 2022, 59, 1509–1533.

Guo C, & Zhu J. Impact of the qingcaosha reservoir project on the sedimentation and erosion of adjacent waters. *Marine Science Research*, 2015, 33(3): 34–41.

Guo J. The formation and evolution of the Shalihong topography outside the Yangtze River Estuary and its role in estuary regulation and development. Shanghai: East China Normal University, 2008.

Guo L, Mick VDW, Jay DA, Matte P, Wang ZB, Roelvink DJA, & He Q. River-tide dynamics: Exploration of nonstationary and nonlinear tidal behavior in the Yangtze River estuary. *Journal of Geophysical Research Oceans*, 2015, 120(5): 3499–3521.

Guo W, Li Y, Zhang L, Zhao R, & Wang H. Analysis of water and sediment changes and driving factors at the Yichang station in the middle reaches of the Yangtze river over the past 60 Years. *Hydrology*, 2019, 39(6): 69–73.

Guo X, Cheng H, Ji N, & Hu H. The impact of the evolution of the Hengsha Channel on the sedimentation of the upper section of the North Channel Deepwater. *Sediment Research*, 2015, (3): 21–26.

Guo X, Cheng H, Yang Z, & Qin Q. Analysis of the evolution and trends of the river regime in the Beigang of the Yangtze river estuary. *Sediment Research*, 2016, (5): 33–39.

Gutierrez RR, Abad JD, Parsons DR, & Best JL. Discrimination of bed form scales using robust spline filters and wavelet transforms: Methods and application to synthetic signals and bed forms of the Río Paraná, Argentina. *Journal of Geophysical Research: Earth Surface*, 2013, 118: 1400–1418.

Guy HP, Simons DB, & Richardson EV. Summary of alluvial channel data from flume experiments, 1956–1961. United States Geological Survey: Reston, VA, USA, 1966. 96.

Ha S, Liu D, & Mu L. Prediction of Yangtze River streamflow based on deep learning neural network with El Niño–Southern Oscil lation. *Scientific Reports*, 2021, 11(1): 1–23.

Haasnoot M, Kwakkel JH, Walker WE, & ter Maat J. Dynamic adaptive policy pathways: A method for crafting robust decisions for a deeply uncertain world. *Global Environmental Change*, 2013, 23(2): 485–498.

Habtu W, & Katihally J. Terrain mapping and analysis for land management: The case of Megech-Dirma watershed, sub-basin of the Blue Nile basin, Northwest Ethiopia. *Arabian Journal of Geosciences*, 2023, 16: 1–27.

Haigh ID, Pickering MD, Green JAM, Arbic BK, Arns A, Dangendorf S, Hill DF, Horsburgh K, Howard T, & Idier D. The tides they are a-Changin': A comprehensive review of past and future nonastronomical changes in tides, their driving mechanisms, and future implications. *Reviews of Geophysics*, 2020, 58(1): 1–39.

Han Y, & Dou X. The course of comprehensive management of the Yangtze River Estuary and reflections. *Ocean Engineering*, 2020(4): 11–18.

Hanasaki N, Fujimori S, Yamamoto T, Yoshikawa S, Masaki Y, Hijioka Y, Kainuma M, Kanamori Y, Masui T, & Takahashi K. A global water scarcity assessment under Shared Socio-economic Pathways – Part 2: water availability and scarcity. *Hydrology and Earth System Sciences*, 2013, 17(7): 2393–2413.

Hao L, Ma N, & He L. Circulation anomaly characteristics of the abnormal drought and high temperature event in the middle and lower reaches of the Yangtze River in summer 2022. *Arid Meteorology*, 2022, 40(5): 721–732.

Hao Z, Zhang X, Hao F, & Fu Y. Research on the impact and response to compound high temperature and drought events—taking the high temperature and drought in the Yangtze river basin in summer 2022 as an example. *Water Resources Protection*, 2023: 1–12.

Haralick, RM Shanmugam, KS, & Dinstein I. Textural features for image classification. *IEEE Trans. Syst. Man Cybern.*, 1973, 3, 610–621.

He L, Pang Y, Zhang Z, Liang X, Chen B. ICESat-2 data classification and estimation of terrain height and canopy height. *International Journal Apply Earth Observation Geoinformation*, 2023, 118, 103233.

He Songlin. Structural pattern of the divergent mouth in the lower reach of the Yangtze river estuary. *Acta Oceanologica Sinica*, 2000, 22(1): 84–92.

He Yufang. Research on the response of the mouth sand bar in the Yangtze river estuary to the relative sea level rise. East China Normal University, 2011.

Herath HMVV, Chadalawada J, & Babovic V. Hydrologically informed machine learning for rainfall–runoff modelling: Towards distributed modelling. *Hydrology and Earth System Sciences*, 2021, 25(8): 4373–4401.

Hobohm C, Schaminee J, & van Rooijen N. Coastal habitats, shallow seas and inland saline steppes: Ecology, distribution, threats and challenges. *Perspectives for Biodiversity and Ecosystems*, 2021, 279–310.

Hochreiter S, Schmidhuber J. Long short-term memory. *Neural Computation*, 1997, 9(8): 1735–1780.

Holben BN. Characteristics of maximum-value composite images from temporal AVHRR data. *International Journal of Remote Sensing*, 1986, 7(11): 1417–1434.

Hou Y, Yin B, Guan C, Guo M, Liu G, & Hu Po. Progress and prospects of marine dynamic disaster research in China. *Oceanology and Limnology*, April 2020 issue.

Jiang Y, Lin L, Chen L, Ni H, Ge W, Cheng H, Zhai G, Wang G, Ban Y, Li Y, Lei M, Tan C, Su J, Zhou Q, Zhang T, Li Y, Liu H, Peng K, & Wang H. Research on conditions of resources and environment and major geological problems in the Yangtze River Economic Zone. *Geology in China*, 2017, 44(6): 1045–1061.

Hu K, & Ding P. The effect of deep waterway constructions on hydrodynamics and salinities in Yangtze Estuary, China. *Journal of Coastal Research*, 2009, 2: 961–965.

Hua K, Cheng H, & Zheng S. Formation mechanism of the nearshore erosional landform in the Hengsha channel of the Yangtze river estuary. *Acta Geographica Sinica*, 2019, 74(7): 1363–1373.

Huang H, Wang Y-P, Gao S, Chen J, Yang Y, & Gao J. Extraction of morphometric bedform characteristics from profiling sonar datasets recorded in shallow coastal waters of China. *China Ocean Engineering*, 2012, 26(3): 469–482.

Huang J. Application of remote sensing and geographic information system technology in the stability evaluation of the river bank in the lower reaches of the Yangtze River. *Geographical Science*, 1999, 19(6): 521–524.

Huang S. Evolution characteristics of the Yangtze River Estuary. *Sediment Research*, 1986, (4): 3–14.

Huang Y, & Lv H. Analysis of the main factors affecting the safety of seawalls and response measures. *China Ocean*, 2009: 838–843.

Huang Z, Yuan X, & Liu X. The key drivers for the changes in global water scarcity: water withdrawal versus water availability. *Journal of Hydrology*, 2021, 601: 126658.

Huang C, Zhang H, & Zhao, J. High-efficiency determination of coastline by combination of tidal level and coastal zone DEM from UAV tilt photogrammetry. *Remote Sensing*, 2020, 12(14), 2189.

Huete AR. A soil-adjusted vegetation index (SAVI). *Remote Sensing of Environment*, 1988, 25, 295–309.

Hunter J. Estimating sea-level extremes under conditions of uncertain sea-level rise. *Climatic Change*, 2010, 99(s3–4): 331–350.

Ibrahim KSMH, Huang YF, Ahmed AN, Koo CH, & El-Shafie A. A review of the hybrid artificial intelligence and optimization modelling of hydrological streamflow forecasting. *Alexandria Engineering Journal*, 2022, 61(1): 279–303.

IPCC. Annex I: Global to Regional Atlas. In: *Climate Change 2022: Impacts, Adaptation and Vulnerability. Contribution of Working Group II to the Sixth Assessment Report of the Intergovernmental Panel on Climate Change*. Cambridge, UK and New York, NY, USA: Cambridge University Press, 2022: 2811–2896.

IPCC Writing Group. IPCC Fifth Assessment Report. 2015.

Iwahashi J, & Pike RJ. Automated classifications of topography from DEMs by an unsupervised nested-means algorithm and a three-part geometric signature. *Geomorphology*, 2007, 86(3): 409–440.

Iwahashi J, Kamiya I, Matsuoka M, & Yamazaki, D. Global terrain classification using 280 m DEMs: Segmentation, clustering, and reclassification. *Progress in Earth and Planetary Science*, 2018, 5, 1–31.

Jabari S, Shahrour I, & El Khattabi J. Assessment of the urban water security in a severe water stress area–application to Palestinian cities. *Water*, 2020, 12(7): 2060.

Jaberzadeh M, Saremi A, Ghorbanizadeh KH, & Babazadeh, H. SWAT and IHACRES models for the simulation of rainfall-runoff of Dez watershed. *Climate Dynamics*, 2022, 1–13.

Jackson L P, Grinsted A, & Jevrejeva S. 21st century sea-level rise in line with the Paris Accord. *Earths Future*, 2018, 6(2): 213–229.

James LA, Watson DG, & Hansen WF. Using LiDAR data to map gullies and headwater streams under forest canopy: South Carolina, USA. *Catena*, 2007, 71(1): 132–144.

Jasiewicz J, Netzel P, & Stepinski TF. Landscape similarity, retrieval, and machine mapping of physiographic units. *Geomorphology*, 2014, 221: 104–112.

Jasiewicz J, & Stepinski TF. Geomorphons—a pattern recognition approach to classification and mapping of landforms. *Geomorphology*, 2013, 182: 147–156.

Jentsje VDM, & Bruce T. New physical insights and design formulas on wave overtopping at sloping and vertical structures. *Journal of Waterway Port Coastal & Ocean Engineering*, 2013, 140(6): 04014025.

Jevrejeva S, Frederikse T, Kopp RE, Le Cozannet G, Jackson LP, van de Wal RSW. Probabilistic sea level projections at the coast by 2100. *Surveys in Geophysics*, 2019, 40(6): 1673–1696.

Ji Na, Cheng Heqin, Yang Zhongyong, Hu Hao, Chen Zujun. Characteristics of sediment and topographic evolution of the Yangtze River Estuary's beach over the past 30 years. *Acta Geographica Sinica*, 2013, 68(7): 945–954.

Ji Y, He G, & Lu Y. Calculation model of the head scour of a typical groyne in the Yangtze River Estuary. *Water Resources and Hydropower Engineering Science Research*, 2000, (4): 43–47.

Ji Y, Lu Y, & Zhang L. Identification of the main influencing factors on seawall safety based on destruction path analysis. *Information Technology in Risk Analysis and Crisis Response—Proceedings of the Sixth Annual Conference of the Risk Analysis Committee of the China Disaster Prevention Association*, 2014.

Jia M, Wang Z, Mao D, Ren C, Wang C, & Wang Y. Rapid, robust, and automated mapping of tidal flats in China using time series Sentinel-2 images and Google Earth Engine. *Remote Sensing of Environment*, 2021, 255, 112285.

Jiang C, Li J, & de Swart HE. Effects of navigational works on morphological changes in the bar area of the Yangtze Estuary. *Geomorphology*, 2012, 140: 205–219.

Jiang Y, Li C, Sun L, Guo D, Zhang Y, & Wang W. A deep learning algorithm for multi-source data fusion to predict water quality of urban sewer networks. *Journal of Cleaner Production*, 2021, 318, 128533.

Jiang Y, Lin L, Chen L, Ni H, Ge W, Cheng H, Zhai G, Wang G, Ban Y, Li Y, Lei M, Tan C, Su J, Zhou Q, Zhang T, Li Y, Liu H, Peng K, & Wang H. Research on conditions of resources and environment and major geological problems in the Yangtze River Economic Zone. *Geology in China*, 2017, 44(6): 1045–1061.

Jiang Z, Cheng H, Hua K, Shi T, Tang M, Teng L, & Yan Ge. Study on scour along the north bank of Hengsha Island in the Yangtze Estuary. *Marine Science Bulletin*, 2020, 39(2): 249–256.

Jiang Z, Zhu D, Shen Y, & Wang X. Numerical simulation method of bank collapse based on theory of critical slip field. *Rock and Soil Mechanics*, 2015, 36(S2): 21–28.

Jin L, Wang N, Fu Q. Analysis of topography of bank-slides and its affecting factors in Mahu reach of the Yangtze River. *Journal of Sediment Research*, 1998, (2): 67–71.

Kang Y, Lei J, Wang M, Li G, & Ding X. Topographic evolution of tidal flats based on remote sensing: An example in Jiangsu coast, Southern Yellow Sea. *Frontiers in Marine Science*, 2023, 10.

Karki R, Qi J, Gonzalez-Benecke CA, Zhang X, Martin TA, & Arnold JG. SWAT-3PG: Improving forest growth simulation with a process-based forest model in SWAT. *Environmental Modelling & Software*, 2023, 164, 105705.

Kelly SF. Water quality in the Yangtze River at Shanghai and the role of coastal reservoirs in Shanghai's water supply. New South Wales: University of Wollongong, 2022.

Kerimoglu O, Große F, Kreus M, Lenhart HJ, van Beusekom JE. A model-based projection of historical state of a coastal ecosystem: Relevance of phytoplankton stoichiometry. *Science of the Total Environment*, 2018, 639: 1311–1323.

Kirwan ML, & Megonigal JP. Tidal wetland stability in the face of human impacts and sea-level rise. *Nature*, 2013, 504(7478): 53–60.

Knaapen MAF, van Bergen Henegouw CN, & Hu YY. Quantifying bed-form migration using multi-beam sonar. *Geo-Marine Letters*, 2005, 25: 306–314.

Knaapen MAF. Sandwave migration predictor based on shape information. *Journal of Geophysical Research*, 2005, 110: F04S11.

Knaapen MAF, & Hulscher SJMH. Regeneration of sand waves after dredging. *Coastal Engineering*, 2002, 46(4): 277–289.

Knaapen MAF, Hulscher SJMH, & de Vriend HJ. A new type of sea bed waves. *Geophysical Research Letters*, 2001, 28(7): 1323–1326.

Kohonen T. *Self Organizing Maps*. 3rd ed. New York: Springer, 2001.

Kostaschuk R. A field study of turbulence and sediment dynamics over subaqueous dunes with flow separation. *Sedimentology*, 2000, 47(3): 519–531.

Kostaschuk RA, & Villard PV. Flow and sediment transport over large subaqueous dunes: Fraser River, Canada. *Sedimentology*, 1996, 43: 849–863.

Kotir JH, Smith C, Brown G, Marshall N, & Johnstone R. A system dynamics simulation model for sustainable water resources management and agricultural development in the Volta River Basin, Ghana. *Science of the Total Environment*, 2016, 573: 444–457.

Kramm T, Hoffmeister D, Curdt C, Maleki S, Khormali F, & Kehl M. Accuracy assessment of landform classification approaches on different spatial scales for the Iranian loess plateau. *ISPRS International Journal of Geo-Information*, 2017, 6(11): 366.

Kratzert F, Klotz D, Shalev G, Klambauer G, & Hochreiter S, Nearing G. Towards learning universal, regional, and local hydro logical behaviors via machine learning applied to large-sample data sets. *Hydrology and Earth System Sciences*, 2019, 23(12): 5089–5110.

Kuang C, Chen W, Gu J, Su TC, Song H, Ma Y, & Dong Z. River discharge contribution to sea-level rise in the Yangtze River Estuary, China. *Continental Shelf Research*, 2017, 134: 63–75.

Kuang C, Liang H, Mao X, Karney B, Gu J, Huang H, Chen W, Song H. Influence of potential future sea-level rise on tides in the China Sea. *Journal of Coastal Research*, 2017, 33(1): 105–117.

Kwadijk JCJ, Haasnoot M, Mulder JPM, Hoogvliet MMC, Jeuken ABM, van der Krogt RAA, van Oostrom NGC, Schelfhout HA, van Velzen EH, van Waveren H, & de Wit MJM. Using adaptation tipping points to prepare for climate change and sea level rise: A case study in the Netherlands. *WIREs Climate Change*, 2010, 1(5): 729–740.

Lacroix P, Berthier E, & Maquerhua ET. Earthquake-driven acceleration of slow-moving landslides in the Colca valley, Peru, detected from Pléiades images. *Remote Sensing of Environment*, 2015, 165: 148–158.

Lacroix P, Bièvre G, Pathier E, Kniess U, Jongmans D. Use of Sentinel-2 images for the detection of precursory motions before landslide failures. *Remote Sensing of Environment*, 2018, 215: 507–516.

Lei J, Li F, Zhang S, Ke H, Zhang Q, Li W. Accuracy assessment of recent global ocean tide models around antarctica. *ISPRS - International Archives of the Photogrammetry, Remote Sensing and Spatial Information Sciences*, 2017, 42: 1521–1528.

Li Bochang. Channel evolution in North Branch of Changjiang River estuary since 1984. *Hydro-Science and Engineering*, 2006, (3): 12–20.

Li B, Yan X X, He Z F, *et al.* Impacts of the Three Gorges Dam on the bathymetric evolution of the Yangtze River Estuary (in Chinese). *Chin Sci Bull*, 2015, 60: 1735–1744.

Li C, Held H, Hokamp S, Marotzke J. Optimal temperature overshoot profile found by limiting global sea level rise as a lower-cost climate target. *Science Advances*, 2020, 6(2): eaaw9490.

Li C, Yang Z, Lu X, Shao L, & Wang J. Bank collapse characteristics, formation mechanism and counterm easures of the Changjiang River in Anhui Province. *Quaternary Sciences*, 2008, 28(4): 578–583.

Li H, Zhao J, Yan B, Yue L, Wang L. Global DEMs vary from one to another: An evaluation of newly released Copernicus, NASA and AW3D30 DEM on selected terrains of China using ICESat-2 altimetry data. *International Journal of Digital Earth*, 2022, 15(1): 1149–1168.

Li J. Research on water resources carrying capacity and coordination in the Yangtze River Delta Region. Shanghai Normal University, 2020.

Li J, Wang Y, Zhang R, Ge Y, Qi D, & Zhang D. Disaster effects of sea level rise—A case of Jiangsu coastal lowland. *Scientia Geographica Sinica*, 2006, 26(1): 87–93.

Li J, Wan X, Ying M, Zhuo S, & Fu G. Study on the formation and evolution of Jiuduan shoal in the Changjiang estuary. *Journal of Sediment Research*, 2006, (6): 44–49.

Li L, Ren Y, Ye T, Wang XH, Hu J, & Xia Y. Positive feedback between the tidal flat variations and sediment dynamics: An example study in the macro-tidal turbid Hangzhou Bay. *Journal of Geophysical Research: Oceans*, 2023, 128(6): 1–25.

Li L, & Zhu J. Impacts of the reclamation project of Nanhui tidal flat on the currents and saltwater intrusion in the Changjiang estuary. *Journal of East China Normal University (Natural Science)*, 2015(4): 77–86.

Li L, Zhu J, Wu H, & Guo Z. Lateral saltwater intrusion in the North Channel of the Changjiang Estuary. *Estuaries & Coasts*, 2014, 37(1): 36–55.

Li M. The coupling interaction for water and sediment discharge and riverbed morphology in the middle and lower Changjiang River channel. East China Normal University, 2005.

Li P, Yang S, Dai S, & Zhang W. Accretion/erosion of the subaqueous delta at the Yangtze Estuary in recent 10 years. *Acta Geographica Sinica*, 2007, 62(7): 707–716.

Li W, & Gong P. Continuous monitoring of coastline dynamics in western Florida with a 30-year time series of Landsat imagery. *Remote Sensing of Environment*, 2016, 179: 196–209.

Li W, Li J, Cheng H, & Liu Q. Research on the dune development in the Changjiang (Yangtze) Estuary China. *Journal of Sediment Research*, 2008, (6): 47–53.

Li W, Niu Z, Shang R, Qin Y, Wang L, Chen H. High-resolution mapping of forest canopy height using machine learning by coupling ICESat-2 LiDAR with Sentinel-1, Sentinel-2 and Landsat-8 data. *International Journal Apply Earth Observation Geoinformation*, 2020, 92, 102163.

Li X, Zhang X, Qiu C, Duan Y, Liu SA, Chen D, Zhang L, & Zhu C. Rapid loss of tidal flats in the Yangtze River Delta since 1974. *International Journal of Environmental Research and Public Health*, 2020, 17(5), 1636.

Li Y, Cai Y, Chen X, Xu Z, & Fu Q. An integrated modeling approach for coupled risk management of regional blue- and greenwater under changing land-use patterns. *Journal of Hydrology*, 2023, 625: 129958.

Li Y, Li Y, & He J. Strategic countermeasures for China's water resources security in the new development stage. *Shuili Xuebao*, 2021, 52(11): 1340–1346, 1354.

Li Z, Cheng H, Lu J, Xiang S, & Teng L. Water shortage and countermeasures of irrigation area around Lake Poyang in dry seasons. *J. Lake Sci.*, 2022, 34(2): 601–615.

Lian W, Zhang G, Cui H, Chen Z, Wei S, Zhu C, & Xie Z. Extraction of high-accuracy control points using ICESat-2 ATL03 in urban areas.

International Journal of Applied Earth Observation and Geoinformation, 2022, 115, 103116.

Liang C, Jaksa MB, Kuo YL, & Ostendorf B. Identifying areas susceptible to high risk of riverbank collapse along the Lower River Murray. *Computers and Geotechnics*, 2015a, 69: 236–246.

Liang C, Jaksa MB, Ostendorf B, & Kuo YL. Influence of river level fluctuations and climate on riverbank stability. *Computers and Geotechnics*, 2015b, 63: 83–98.

Liang W, Liu T, Wang Y, Jiao JJ, Gan J, He D. Spatiotemporal-aware machine learning approaches for dissolved oxygen prediction in coastal waters. *Science of the Total Environment*, 2023, 905, 167138.

Liao WH. Region description using extended local ternary patterns, August 23, 2010. IEEE, 2010.

Liao X, & Wang K. Study on application of fuzzy statistic clustering theory in bank-failure. *Journal of Yangtze River Scientific Research Institute*, 2007, 24(2): 5–8.

Lin CW, Tseng CM, Tseng YH, Fei LY, Hsieh YC, Tarolli P. Recognition of large scale deep-seated landslides in forest areas of Taiwan using high resolution topography. *Journal of Asian Earth Sciences*, 2013, 62: 389–400.

Lin M, Lu J, Zhang D, Chen H, Zhang Z, & Gao H. Evolution and regulation works at Hechangzhou braided channel in Z-Y Reach of Yangtze River. *Journal of Yangtze River Scientific Research Institute*, 2006, 23(5): 10–13.

Liu B, Qin X, Zhang F. System-dynamics-based scenario simulation and prediction of water carrying capacity for China. *Sustainable Cities and Society*, 2022, 82: 103912.

Liu D, Ye Y, Li D, & Chen X. GIS-supported analysis on riverbed evolution in lower segment of Southern Channel in Yangtze Estuary. *Coastal Engineering*, 2010, 29(3): 19–27.

Liu G, Shen H, Wu J, & Wu H. The study of hydrodynamic characteristics of flood and ebb channels and channel type judgment. *Acta Oceanologica Sinica*, 2005, 27(5): 151–156.

Liu G, Gong Y, Wang H & Lai X. Current situation of levee risk management abroad and its implications to China. *Journal of Yangtze River Scientific Research Institute*, 2019, 36(10): 53–58.

Liu GW, Cheng HQ, Li JF, Ji N, Hu H, & Qiao YY. Variations in tidal currents and suspended sediment concentration of the upper part of the North Channel of Changjiang Estuary over the past 10 years. *Marine Science Bulletin*, 2015, 17(2): 428–435.

Liu J, Cheng H, Han L & Wang Z. Influence of fluvial sediment decline on the morphdynamics of the Yangtze River Estuary and adjacent seas. *Advances in Water Science*, 2017, (2): 92–99.

Liu J, Zhao D, & Chen H. Recent morphological evolution of Jiuduan Shoal and its effect on navigation channel sedimentation of North Passage in Yangtze River Estuary. *Journal of Yangtze River Scientific Research Institute*, 2010, 27(7): 5–9.

Liu L, Wang YQ, Zhan RF, Xu J, Duan YH. Increasing destructive potential of landfalling tropical cyclones over China. *Journal of Climate*, 2020, 33(9): 3731–3743.

Liu M, Wei J, Wang G, & Wang F. Water resources stress assessment and risk early warning–a case of Hebei Province China. *Ecological Indicators*, 2017, 73: 358–368.

Liu Y, & Wu L. Geological disaster recognition on optical remote sensing images using deep learning. *Procedia Computer Science*, 2016, 91: 566–575.

Liu A, Cheng X, & Chen Z. Performance evaluation of GEDI and ICESat-2 laser altimeter data for terrain and canopy height retrievals. *Remote Sensing of Environment*, 2021, 264, 112571.

Liu HQ, & Huete AA feedback based modification of the NDVI to minimize canopy background and atmospheric noise. *IEEE Transactions on Geoscience and Remote Sensing*, 1995, 33(2), 457–465.

Liu, X. Airborne LiDAR for DEM generation: Some critical issues. *Progress in Physical Geography*, 2008, 32, 31–49.

Liu X, Lin M, Li Z. Channel evolution and regulation of Zhenjiang-Yangzhou reach of the lower Yangtze River. *Journal of Yangtze River Scientific Research Institute*, 2011, (11): 1–9.

Liu X, Wang X, Guan X, Kong L, & Su A. Study on storage capacity and characteristic water level for Qingcaosha Reservoir. *Water Resources and Hydropower Engineering*, 2009, 40(7): 5–8, 48.

Liu X, Zhang Z, Peterson J, & Chandra S. The effect of LiDAR data density on DEM accuracy. 2007.

Liu X, Zhong C, Wang Z, Zhang X, Wang J, Cai Q, Chen Y, Zhu S, & Yu X. Distribution characteristics of oily contaminant in Wusong Estuary sediments. *Journal of Shanghai University (Natural Science)*, 2010, (6): 592–596.

Liu X, & Zhao Y. Estimation on indirect economic losses of geomorphic hazards_Taking debris flow as an example. *Progress in Geography*, 2008, 27(3): 7–12.

Liu Z, Ding B, Guan S, Chen J, Zhang Y & Yu H. Numerical simulation and analysis of saltwater intrusion to Modaomen waterway. *Acta Scientiaum Naturalium Universitatis Sunyatseni*, 2016, 55(6): 1–9.

Lu X, Cheng H, Zhou Q *et al.* Features and mechanism of asymmetric double-kidneys sourced geomorphology of pier in tidal estuary. *Haiyang Xuebao*, 2016, 38(9): 118–125.

Lu X, Li Y, Washington-Allen RA, Li Y, Li H, & Hu Q. The effect of grid size on the quantification of erosion, deposition, and rill network. *International Soil and Water Conservation Research*, 2017, 5: 241–251.

Lu Y, He Y, & Liu H. Research on seawall flood defense criteria. *Engineering Science*, 2005, 7(12): 17–23.

Luan HL, Ding PX, Wang ZB, & Ge JZ. Process-based morphodynamic modeling of the Yangtze Estuary at a decadal timescale: Controls on estuarine evolution and future trends. *Geomorphology*, 2017, 290(8): 347–364.

Luan HL, Ding PX, Wang ZB, Ge JZ, & Yang SL. Decadal morphological evolution of the Yangtze Estuary in response to river input changes and estuarine engineering projects. *Geomorphology*, 2016, 265: 12–23.

Luan H, Ke K, Ge J, Chen L, Yang W, & Ding P. 3D numerical simulation of the impacts of planned estuarine projects on saltwater intrusion in the Yangtze Estuary. *Advances in Marine Science*, 2018, 36(4): 525–539.

Luan H, Liu T, & Huang W. Morphological evolution and trends of typical central bar channels in the lower Yangtze River under varying water and sediment discharge. *Journal of Yangtze River Scientific Research Institute*, 2018, 35(11): 11–16.

Luchi R, Hooke JM, & Zolezzi G. Width variations and mid-channel bar inception in meanders: River Bollin(UK). *Geomorphology*, 2010, 119(1–2): 1–8.

Luo W, & Liu C- C. Innovative landslide susceptibility mapping supported by geomorphon and geographical detector methods. *Landslides*, 2018, 15(3): 465–474.

Ma G, Shi F, Liu S, & Qi D. Hydrodynamic modeling of Changjiang Estuary: Model skill assessment and large-scale structure impacts. *Applied Ocean Research*, 2011, 33(1): 69–78.

Ma ZJ, Melville DS, Liu JG, Chen Y, Yang HY, Ren WW, Zhang ZW, Piersma T, & Li B. Rethinking China's new great wall. *Science*, 2014, 346(6212): 912–914.

Ma S, Wang N, Zhou L, Yu J, Chen X, & Chen Y. Inversion of tidal flat topography based on the optimised inundation frequency method—a case study of intertidal zone in Haizhou Bay, China. *Remote Sensing*, 2024, 16(4): 685–685.

MacMillan RA, Pettapiece WW, Nolan SC, & Goddard TW. A generic procedure for automatically segmenting landforms into landform elements using DEMs, heuristic rules and fuzzy logic. *Fuzzy Set and Systems*, 2000, 113(1): 81–109.

Malambo L, & Popescu SC. Assessing the agreement of ICESat-2 terrain and canopy height with airborne lidar over US ecozones. *Remote Sensing of Environment*, 2021, 266: 112711. DOI:10.1016/j.rse.2021.112711.

Mancini F, & Dubbini M. *Unmanned Aerial System Applications to Coastal Environments*, 2020, 87–109.

Mao X, Gu S, & Tang G. Impact of tidal level change on tidal form in Yangtze Estuary. *Journal of China Hydrology*, 2013, 33(2): 11–14.

Marcel Van RAG, & Ivo MVDW. Rock toe stability of rubble mound breakwaters. *Coastal Engineering*, 2014, 83(1): 166–176.

Maritime works - Part 1-2: General -Code of practice for assessment of actions BS 6349-1-2: 2016.

Markus T, Neumann T, Martino A, Abdalati W, Brunt K, Csatho B, Farrell S, Fricker H, Gardner A, Harding D, & Jasinski M. The Ice, Cloud, and land Elevation Satellite-2 (ICESat-2): Science requirements, concept, and implementation. *Remote Sensing of Environment*, 2017, 190, 260–273.

Markus T, Neumann TA, Martino AJ, Abdalati W, Brunt KM, Csathó BM, Farrell SL, Fricker HA, Gardner AS, Harding DJ, Jasinski MF, Kwok R, Magruder LA, Lubin D, Luthcke SB, Morison J, Nelson RF, Neuenschwander AL, Palm SP, Popescu SC, Shum CK, Schutz BE, Smith BE, Yang Y, Zwally JH. The Ice, Cloud, and land Elevation Satellite-2 (ICESat-2): Science requirements, concept, and implementation. *Remote Sensing of Environment*, 2017, 190, 260–273.

Martha TR, Kerle N, van Westen CJ, Jetten V, & Kumar KV. Object-oriented analysis of multi-temporal panchromatic images for creation of historical landslide inventories. *ISPRS Journal of Photogrammetry and Remote Sensing*, 2012, 67: 105–119.

Mason DC, Davenport IJ, Robinson GJ, Flather RA & McCartney BS. Construction of an inter-tidal digital elevation model by the 'Water-Line' Method. *Geophysical Research Letters*, 1995, 22(23), 3187–3190.

McAdoo BG, Pratson LF, & Orange DL. Submarine landslide geomorphology, US continental slope. *Marine Geology*, 2000, 169(1–2): 103–136.

Mckay DIA, Staal A, Abrams JF, Winkelmann R, Sakschewski B, Loriani S, Fetzer I, Cornell SE, Rockström J, & Lenton TM. Exceeding 1.5°C global warming could trigger multiple climate tipping points. *Science*, 2022, 377(6611): 1171–1171.

Mcleod E, Poulter B, Hinkel J, Reyes E, Salm R. Sea-level rise impact models and environmental conservation: A review of models and their applications. *Ocean & Coastal Management*, 2010, 53(9): 507–517.

Meehl GA, Hu A, Tebaldi C, Arblaster JM, Washington WM, Teng H, Sanderson BM, Ault T, Strand WG, & White JB. Relative outcomes of climate change mitigation related to global temperature versus sea-level rise. *Nature Climate Change*, 2012, 2(8): 576–580.

Meyers SD, Linville AJ, & Luther ME. Alteration of residual circulation due to large-scale infrastructure in a coastal plain estuary. *Estuaries and Coasts*, 2014, 37: 493–507.

Miliaresis GC, & Argialas DP. Segmentation of physiographic features from the global digital elevation model/GTOPO30. *Computers & Geosciences*, 1999, 25(7): 715–728.

Minár J, & Evans IS. Elementary forms for land surface segmentation: The theoretical basis of terrain analysis and geomorphological mapping. *Geomorphology*, 2008, 95(3–4): 236–259.

Mo L, Lou S, Wang Y, Liu Z, & Ren P. Studying the evolutions, differences, and water security impacts of water demands under shared socioeconomic pathways: A SEMs-bootstrap-ANN approach applied to Sichuan Province. *Journal of Environmental Management*, 2024, 349: 119455.

Mo R, Guo X, & Yang Z. Effect of large-scale hydraulic projects on river regime evolution in North Channel of Yangtze estuary. *Port & Waterway Engineering*, 2015, (12): 98–103.

Mockett ID, Simm JD. *Risk Levels in Coastal and River Engineering.* Thomas Telford, 2002.

Moftakhari HR, Salvadori G, AghaKouchak A, Sanders BF, & Matthew RA. Compounding effects of sea level rise and fluvial flooding. *Proceedings of the National Academy of Sciences*, 2017, 114(37): 9785–9790.

Mohammadi A, Karimzadeh S, Jalal SJ, Kamran KV, Shahabi H, Homayouni S, & Al-Ansari N. A multi-sensor comparative analysis on the suitability of generated DEM from Sentinel-1 SAR interferometry using statistical and hydrological models. *Sensors (Basel, Switzerland)*, 2020, 20.

Mondini AC, Guzzetti F, Reichenbach P, Rossi M, Cardinali M, & Ardizzone F. Semi-automatic recognition and mapping of rainfall induced shallow landslides using optical satellite images. *Remote Sensing of Environment*, 2008, 115(7): 1743–1757.

Moosavi V, Talebi A, & Shirmohammadi B. Producing a landslide inventory map using pixel-based and object-oriented approaches optimized by Taguchi method. *Geomorphology*, 2014, 204(1): 646–656.

Morelissen R, Hulscher SJMH, Knaapen MAF, Nemeth AA, & Bijker R. Mathematical modelling of sand wave migration and the interaction with pipelines. *Coastal Engineering*, 2003, 48(3): 197–209.

Mudersbach C, Wahl T, Haigh ID, & Jensen J. Trends in high sea levels of German North Sea gauges compared to regional mean sea level changes. *Continental Shelf Research*, 2013, 65: 111–120.

Murray NJ, Worthington TA, Bunting P, Duce S, Hagger V, Lovelock CE, Lucas R, Saunders MI, Sheaves M, Spalding M, Waltham NJ. High-resolution mapping of losses and gains of Earth's tidal wetlands. *Science*, 2022, 376(6594), 744–749.

Nandy S, Srinet R, Padalia H. Mapping forest height and aboveground biomass by integrating ICESat-2, Sentinel-1 and Sentinel-2 data using random forest algorithm in Northwest Himalayan foothills of India. *Geophysical Research Letters*, 2021, 48(14): 1–10.

Nasrollahi M, Khosravi H, Moghaddamnia A, Malekian A, Shahid S. Assessment of drought risk index using drought hazard and vulnerability indices. *Arabian Journal of Geosciences*, 2018, 11(606): 1–12.

Nauels A, Gütschow J, Mengel M, Meinshausen M, & Schleussner CF. Attributing long-term sea-level rise to Paris Agreement emission pledges. *Proceedings of the National Academy of Sciences*, 2019, 116(47): 23487–23492.

Nelson A, Reuter HI, & Gessler P. Chapter 3 dem production methods and sources. *Developments in Soil Science*, 2009, 33, 65–85.

Neuenschwander AL, Guenther E, White JC, Duncanson LI, & Montesano PM. Validation of ICESat-2 terrain and canopy heights in boreal forests. *Remote Sensing of Environment*, 2020, 251, 112110.

Neuenschwander AL, & Pitts KL. The ATL08 land and vegetation product for the ICESat-2 Mission. *Remote Sensing of Environment*, 2019, 221: 247–259.

Neumann TA, Brenner A, Hancock D, Robbins J, Saba J, Harbeck K, & Rebold T. ATLAS/ICESat-2 L2A global geolocated photon data, version 3. Boulder, CO: NASA National Snow and Ice Data Center Distributed Active Archive Center. Retrieved from, 2020, 10.

Neumann TA, Martino AJ, Markus T, Bae S, Bock MR, Brenner AC, Brunt KM, Cavanaugh J, Fernandes ST, Hancock DW, Harbeck K, Lee J, Kurtz NT, Luers PJ, Luthcke SB, Magruder LA, Pennington TA, Ramos-Izquierdo L, Rebold TW, Skoog J, & Thomas TC. The Ice, Cloud, and Land Elevation Satellite - 2 Mission: A global geolocated photon product derived from the Advanced Topographic Laser Altimeter System. *Remote Sensing of Environment*, 2019, 233.

Noori N, Kalin L, & Isik S. Water quality prediction using SWAT-ANN coupled approach. *Journal of Hydrology*, 2020, 590, 125220.

Noto LV, Bastola S, Dialynas YG, Arnone E, & Bras RL. Integration of fuzzy logic and image analysis for the detection of gullies in the Calhoun Critical Zone Observatory using airborne LiDAR data. *ISPRS Journal of Photogrammetry and Remote Sensing*, 2017, 126: 209–224.

Novellino A, Cigna F, Sowter A, Ramondini M, & Calcaterra D. Exploitation of the Intermittent SBAS (ISBAS) algorithm with COSMO-SkyMed data for landslide inventory mapping in north-western Sicily, Italy. *Geomorphology*, 2017, 280: 153–166.

Omer A, Elagib NA, Zhuguo M, Saleem F, & Mohammed A. Water scarcity in the Yellow River Basin under future climate change and human activities. *Science of the Total Environment*, 2020, 749: 141446.

Oppenheimer M, Glavovic BC, Hinkel J, Roderik VDW, Magnan A, Abd-Elgawad A, Cai RS, Miguel CJ, Robert, DC, Ghosh T, Hay J, Marzeion B, Meyssignac B, Sebesvari Z, Smit AJ, Dangendorf S, & Frederikse T. Sea level rise and implications for low-lying islands, coasts and communities. In: IPCC Special Report on the Ocean and Cryosphere in a Changing Climate [H.-O. Pörtner, *et al.* (eds.)]. 2019.

Ouyang S, Xu C, Shao J, & Hu F. Preliminary study on water storage situation of cascade reservoirs in the upstream of Yangtze River under drought condition. *Journal of Yangtze River Scientific Research Institute*, 2023: 1–9.

Ortuño M, Guinau M, Calvet J, Furdada G, Bordonau J, Ruiz A, & Camafort M. Potential of airborne LiDAR data analysis to detect subtle landforms of slope failure: Portainé, Central Pyrenees. *Geomorphology*, 2017, 295: 364–382.

Pan JN, Wang P, Sun TT, Chen MW, & Wang DT. Experimental study on inner slope failure mechanism of seawall by coupling effect of storm surge and wave. *Journal of Oceanology and Limnology*, 2019, 37(6): 1912–1920.

Pan J, Wang D, Wu M, & Yang Z. Experimental study on stability of concrete block revetment under wave action. *Journal of Hohai University Natural Sciences*, 2005(4): 476–481.

Pan L, Ding P, Ge J & Hu K. Analysis of influence of Deep Waterway Project on morphological change in North Passage of Changjiang Estuary. *Journal of Sediment Research*, 2011, (5): 53–61.

Parsons DR, Best JL, Orfeo O, Hardy RJ, Kostaschuk R, & Lane SN. Morphology and flow fields of three-dimensional dunes, Rio Parana, Argentina: Results from simultaneous multibeam echo sounding and acoustic Doppler current profiling. *Journal of Geophysical Research*, 2005, 110: 1–9.

Passalacqua P, Belmont P, Foufoula-Georgiou E. Automatic geomorphic feature extraction from lidar in flat and engineered landscapes. *Water Resources Research*, 2012, 48(3): L06.

Peng Y, Xiong C, & Yang C. Analysis of relationship between fluvial process and bank caving in the Jingjiang reach of Yangtze River. *Journal of China Hydrology*, 2010, 30(6): 29–36.

Pinto LC, de Mello CR, Norton LD, Owens PR, & Curi N. Spatial prediction of soil–water transmissivity based on fuzzy logic in a Brazilian headwater watershed. *Catena*, 2016, 143: 26–34.

Pokhrel Y, Felfelani F, Satoh Y, Boulange J, Burek P, Gädeke A, Gerten D, Gosling SN, Grillakis M, Gudmundsson L, Hanasaki N, Kim H, Koutroulis A, Liu J, Papadimitriou L, Schewe J, Müller SH, Stacke T, Telteu C, Thiery W, Veldkamp T, Zhao F, & Wada Y. Global terrestrial water storage and drought severity under climate change. *Nature Climate Change*, 2021, 11(3): 226–233.

Potapov PV, Li X, Hernández-Serna A, Tyukavina A, Hansen MC, Kommareddy A, Pickens AH, Turubanova S, Tang H, Silva CE, Armston J, Dubayah RO, Blair JB, & Hofton MA. Mapping global forest canopy height through integration of GEDI and Landsat data. *Remote Sensing of Environment*, 2020, 112165.

Prima ODA, Echigo A, Yokoyama R, Yoshida T. Supervised landform classification of Northeast Honshu from DEM-derived thematic maps. *Geomorphology*, 2006, 78(3): 373–386.

Pritchard DW. The dynamic structure of a coastal plain estuary. *Journal of Marine Research*, 1956, 15: 33–42.

Pritchard DW. What is an estuary: Physical viewpoint. American Association for the Advancement of Science, 1967, 149–176.

Pu X, Shi JZ, Hu GD, Xiong LB. Circulation and mixing along the North Passage in the Changjiang River estuary, China. *Journal of Marine Systems*, 2015, 148: 213–235.

Puga-Bernabéu Á, Webster JM, Beaman RJ, & Guilbaud V. Morphology and controls on the evolution of a mixed carbonate–siliciclastic submarine canyon system, Great Barrier Reef margin, north-eastern Australia. *Marine Geology*, 2011, 289(1–4): 100–116.

Pugh DJ, & Vassie JM. Extreme sea-levels from tide and surge probability. *Coastal Engineering Proceedings*, 1978, 1(16): 52.

Qiu C, & Zhu J. Impact of sea-level rise on saltwater intrusion in the Changjiang Estuary in the low river discharge cases. *Progressus Inquisitionses De Mutatione Climates*, 2015, 11(4): 245–255.

Rahman MF, Mukherji A, Johannessen A, Srivastava S, Verhagen J, Ovink H, Ligtvoet W, & Olet E. As the UN meets, make water central to climate action. *Nature*, 2023, 615(7953): 582–585.

Ranndal H, Christiansen PS, Kliving P, Andersen OB, Nielsen K. Evaluation of a statistical approach for extracting shallow water bathymetry signals from ICESat-2 ATL03 photon data. *Remote Sensing*, 2021, 13, 3548.

Rasmussen DJ, Bittermann K, Buchanan MK, Kulp S, Strauss BH, & Kopp R, Oppenheimer M. Extreme sea level implications of 1.5°C,

2.0°C, and 2.5°C temperature stabilization targets in the 21st and 22nd century. *Environmental Research Letters*, 2018, 13: 1–12.

Ray BR, & Rahman A. Development and application of regional urban water security indicators. *Utilities Policy*, 2023, 84: 101637.

Razak KA, Santangelo M, Van Westen CJ, Straatsma MW, & de Jong SM. Generating an optimal DTM from airborne laser scanning data for landslide mapping in a tropical forest environment. *Geomorphology*, 2013, 190: 112–125.

Rodríguez JF, Saco PM, Sandi S, Saintilan N, & Riccardi G. Potential increase in coastal wetland vulnerability to sea-level rise suggested by considering hydrodynamic attenuation effects. *Nature Communications*, 2017, 8(1), 16094.

Romstad B, & Etzelmüller B. Mean-curvature watersheds: A simple method for segmentation of a digital elevation model into terrain units. *Geomorphology*, 2012, 139–140: 293–302.

Rouse JW, Haas RH, Schell JA, & Deering DW. Monitoring vegetation systems in the Great Plains with ERTS. 1974, 1, 309–317.

Ruan W, Cao H, & Gong H. The study of the north-south channel bifurcation control system and its effect. *The Ocean Engineering*, 2011, 29(3): 76–81.

Rubin DM. A unifying model for planform straightness of ripples and dunes in air and water. *Earth-Science Reviews*, 2012, 113(3–4): 176–185.

Salvatierra MM, Aliotta S, & Ginsberg SS. Morphology and dynamics of large subtidal dunes in Bahia Blanca estuary, Argentina. *Geomorphology*, 2015, 246: 168–177.

Schaeffer M, Hare W, Rahmstorf S, & Vermeer M. Long-term sea-level rise implied by 1.5°C and 2°C warming levels. *Nature Climate Change*, 2012, 2(8): 867–870.

Scully ME, & Geyer WR. The role of advection, straining, and mixing on the tidal variability of estuarine stratification. *Journal of Physical Oceanography*, 2012, 42(5): 855–868.

Seminara G, Zolezzi G, & Tubino M. Downstream and upstream influence in river meandering. Part 2. Planimetric development. *Journal of Fluid Mechanics*, 2001, 438: 213–230.

Shao C. Lateral secondary flow, mixing and stratification within the curved channel of the North Passage in the Changjiang River Estuary: Data analysis and mathematical modeling. Shanghai Jiaotong University, 2016.

Shi S, Cheng H, Zheng S, *et al.* Erosional topography of the tidal limit in the Yangtze River in flood seasons after the river closure at Three Gorges. *Haiyang Xuebao*, 2017, 39(3): 85–95.

Shi SY, Cheng HQ, Xuan XN, Jiang YH, & Zhou QP. Fluctuations in the tidal limit of the Yangtze River estuary in the last decade. *Science China Earth Sciences*, 2018, 61: 1136–1147.

Shi S. Recent change of the tidal limit and its adaptive channel morphology of the Changjiang (Yangtze) River Estuary. East China Normal University, 2017.

Shi T, Cheng H, Hua K, Teng L, Tang M, Jiang Z, Yan G, Jiang Y, & Zhou Q. Recent evolution of flood channels in the Yangtze Estuary affected by fluvial and estuarine engineering: An instance of the Liuyao flood channel. *Marine Science Bulletin*, 2020, 39(1): 134–142.

Shi X, Tan J, Guo Z *et al.* A review of risk assessment of storm surge disaster. *Advances in Earth Science*, 2013, 28(8): 866–874.

Shih SS, Hsu WC, & Hsu YW. Waterline digital elevation model development to quantify inundation duration and coastal protection of tidal wetlands. *Science of the Total Environment*, 2023, 874, 162519.

Silva CA, Duncanson LI, Hancock S, Neuenschwander AL, Thomas NM, Hofton MA, Fatoyinbo L, Simard M, Marshak C, Armston J, Lutchke SB, & Dubayah RO. Fusing simulated GEDI, ICESat-2 and NISAR data for regional aboveground biomass mapping. *Remote Sensing of Environment*, 2021, 253(112234): 1–14.

Simard M, Pinto N, Fisher JB, Baccini A. Mapping forest canopy height globally with spaceborne lidar. *Journal of Geophysical Research*, 2011, 116.

Simarro G, Guillen J, Puig P, Ribo M, Iacono CL, Palanques A, Munoze A, Durana R, & Acosta A. Sediment dynamics over sand ridges on a tideless mid-outer continental shelf. *Marine Geology*, 2015, 361: 25–40.

Simons DB, Richardson EV. Resistance to flow in alluvial channels. United States Geological Survey: Reston, VA, USA, 1966.

Simons DB, Richardson EV, & Albertson ML. Flume studies using medium sand (0.45 mm). United States Geological Survey: Reston, VA, USA, 1961.

Simpson JH, Brown J, & Allen JM. Tidal straining, density currents, and stirring in the control of estuarine stratification. *Estuaries*, 1990, 13(2): 125–132.

Singh A, Fienberg K, Jerolmack DJ, Marr J, Foufoula-Georgiou E. Experimental evidence for statistical scaling and intermittent in sediment transport rate. *Journal of Geophysical Research*, 2009, 114: F01025.

Smith RL. Extreme value theory based on the r largest annual events. *Journal of Hydrology*, 1986, 86(1–2): 27–43.

So JM, Lee JH, & Bae DH. Development of a hydrological drought forecasting model using weather forecasting data from GloSea5. *Water*, 2020, 12(10), 2785.

Song XM, Kong FZ, Zhan CS, & Han JW. Hybrid optimization rainfall-runoff simulation based on Xinanjiang model and artifi cial neural network. *Journal of Hydrologic Engineering*, 2012, 17(9), 1033–1041.

Soulsby RL, Atkins R, Water CB, Oliver N. Field measurements of suspended sediment over sandwaves. In: Sand Transport in Rivers, Estuarine and the Sea (Ed. By Soulsby R L and Bettess R). Balkema, Rotterdam, 1991, 155–162.

Starek MJ, Mitásová H, Hardin E, Weaver K. Overton M, & Harmon RS. Modeling and analysis of landscape evolution using airborne, terrestrial, and laboratory laser scanning. *Geosphere*, 2011, 7, 1340–1356.

Stefano MD, & Mayer LA. An automatic procedure for the quantitative characterization of submarine bedforms. *Geosciences*, 2018, 8(1): 28.

Stumpf A, & Kerle N. Combining random forests and object-oriented analysis for landslide mapping from very high resolution imagery. *Procedia Environmental Sciences*, 2011b, 3(1): 123–129.

Stumpf A, & Kerle N. Object-oriented mapping of landslides using random forests. *Remote Sensing of Environment*, 2011a, 115(10): 2564–2577.

Sun Yi. Accretion-erosion evolution and its driving mechanisms of Yangtze underwater delta over the past nearly 20 years. Chengdu University of Technology, 2018.

Sun Yi, Li Bing, Yan X & Li Y. Effect of North-South Channel bifurcation control engineering on the North Channel of Yangtze Estuary. *Journal of Sediment Research*, 2018, 43(5): 33–38.

Sun J, Cheng Z, Yang R, & Shan S. Soft-sensor modeling for paper mill effluent COD based on PCA-PSO-LSSVM. *Computers and Applied Chemistry*, 2017, 34(09): 706–710.DOI:10.16866/j.com.app.chem201709010.

Susnik J, Vamvakeridou-Lyroudia LS, Savić DA, Kapelan Z. Integrated system dynamics modelling for water scarcity assessment: Case study of the Kairouan region. *Science of the Total Environment*, 2012, 440: 290–306.

Swatantran A, Tang H, Barrett T, DeCola PL, Dubayah RO. Rapid, high-resolution forest structure and terrain mapping over large areas using single photon Lidar. *Scientific Reports*, 2016, 6.

Syvistski JPM, Albert JK, Irina O, Hutton EW, Hannon MT, Brakenridge GR, Day J, Vorosmarty C, Saito Y, & Giosan L. Sinking deltas due to human activities. *Nature Geosciences*, 2009(2): 681–686.

Talke SA, & Jay DA. Changing tides: The role of natural and anthropogenic factors. *Annual Review of Marine Science*, 2020, 12(1): 121–151.

Tan K, Chen J, Zhang W, Liu K, Tao P, & Cheng X. Estimation of soil surface water contents for intertidal mudflats using a near-infrared long-range terrestrial laser scanner. *ISPRS Journal of Photogrammetry and Remote Sensing*, 2020, 159, 129–139.

Tan R, Wang Z, Wu T, & Wu J. A data-driven model for water quality prediction in Tai Lake, China, using secondary modal decomposition with multidimensional external features. *Journal of Hydrology: Regional Studies*, 2023, 47, 101435.

Tang J, Deng J, You X, & Wang F. Forecast method for bank collapse in middle and lower Yangtze River (MLYZ). *Journal of Sichuan University (Engineering Science Edition)*, 2012, 44(1): 75–81.

Tang M, Li Y, Wang W, Xu W, Guan Z, Wen Z, Guan X. Risk assessment method of level-induced water scarcity in the city located along rivers: Application in Nanchang, China. *Ecological Indicators*, 2023, 157: 111256.

Tang Y, Xu G, Xu H, & Ying R. Study on clean water diversion tests of river network in Chongming Island. *Water Resources Projection*, 2011, 27(4): 70–73.

Tao J, Qi H, Cao L, & Zhang C. Analysis on the impact of runoff variation on saltwater intrusion in Yangtze River Estuary. *China Science Paper*, 2014(11): 1233–1236.

Taveneau A, Almar R, Bergsma EW, Sy BA, Ndour A, Sadio M, Garlan T. Observing and predicting coastal erosion at the Langue de Barbarie sand spit around Saint Louis (Senegal, West Africa) through satellite-derived digital elevation model and shoreline. *Remote Sensing*, 2021, 13(13), 2454.

Tawn JA. An extreme-value theory model for dependent observations. *Journal of Hydrology*, 1988, 101(1): 227–250.

Tawn JA. Estimating probabilities of extreme sea-levels. *Journal of the Royal Statistical Society*, 1992, 41(1): 77–93.

Tawn JA, & Vassie JM. Extreme sea-levels: The joint probabilities method revisited and revised. *Proceedings of the Institute of Civil Engineering Part*, 1989, 2(87): 249–442.

Teng L, Cheng H, & Xu W, Zheng Shuwei. Study on scour along the levee of the deep-water channel project in Yangtze Estuary. *Journal of Sediment Research*, 2019, 44(4): 41–46.

Tian Z, Ramsbottom D, Sun L, Huang Y, Zou H, & Liu J. Dynamic adaptive engineering pathways for mitigating flood risks in Shanghai with regret theory. *Nature Water*, 2023, 1(2): 198–208.

Tian Z, Wu W, Liu J, & Su L. A review of decision-making methods for climate change adaptation under deep uncertainty: with a focus on flooding control in coastal cities. *China Science Bullinten*, 2022, 67: 2638–2650.

Tian X, & Shan J. Comprehensive evaluation of the ICESat-2 ATL08 terrain product. *IEEE Transactions on Geoscience and Remote Sensing*, 2021, 59, 8195–8209.

Tsai YLS, & Tseng KH. Monitoring multidecadal coastline change and reconstructing tidal flat topography. *International Journal of Applied Earth Observation and Geoinformation*, 2023, 118, 103260.

Turpie KR, Klemas VV, Byrd K, Kelly M, & Jo YH. Prospective HyspIRI global observations of tidal wetlands. *Remote Sensing of Environment*, 2015, 167, 206–217.

Twichell DC, Chaytor JD, ten Brink US, & Buczkowski B. Morphology of late quaternary submarine landslides along the US Atlantic continental margin. *Marine Geology*, 2009, 264(1–2): 4–15.

Urbazaev M, Hess LL, Hancock S, Sato LY, Ometto J P, Thiel C, Dubois C, Heckel K, Urban M, Adam M, & Schmullius C. Assessment of terrain elevation estimates from ICESat-2 and GEDI spaceborne LiDAR missions across different land cover and forest types. *Science of Remote Sensing*, 2022, 6, 100067.

Valle-Levinson A. Density-driven exchange flow in terms of the Kelvin and Ekman numbers. *Journal of Geophysical Research Atmospheres*, 2008, 113(C4): 1–10.

Veldkamp T, Zhao F, Ward PJ, Moel HD, Aerts J, Schmied HM, Portmann FT, Masaki Y, Pokhrel YN, & Liu X. Human impact parameterizations in global hydrological models improve estimates of monthly discharges and hydrological extremes: A multi-model validation study. *Environmental Research Letters*, 2018, 13(055008): 1–17.

Vesanto J, & Alhoniemi E. Clustering of the self-organizing map. *IEEE Transactions on Neural Networks*, 2000, 11(3): 586–600.

Vinci A, Brigante R, Todisco F, Mannocchi F, & Radicioni F. Measuring rill erosion by laser scanning. *Catena*, 2015, 124: 97–108.

Wai KP, Chia MY, Koo CH, Huang YF, & Chong WC. Applications of deep learning in water quality management: A state-of-the-art review. *Journal of Hydrology*, 2022, 613, 128332.

Wanders N, Thober S, Kumar R, Pan M, Sheffield J, Samaniego L, & Wood EF. Development and evaluation of a pan-European multimodel seasonal hydrological forecasting system. *Journal of Hydrometeorology*, 2019, 20(1), 99–115.

Wang H, Ge Z, Yuan L, & Zhang L. Evaluation of the combined threat from sea-level rise and sedimentation reduction to the coastal wetlands in the Yangtze Estuary, China. *Ecological Engineering*, 2014, 71, 346–354.

Wang H, Huang J, Cheng X, Zhou H, & Yuan Y. Scenario simulation of water resources development and utilization based on a system dynamics model. *International Journal of Water Resources Development*, 2022, 38(3): 447–463.

Wang H, Liu H, & Yang F. Numerical simulation of wave propagation and wave-breaking on slope topography. *The Ocean Engineering*, 2020, 38(4): 54–60.

Wang H, Qian L, & Zhao Z, WANG Ying. Theory and assessment method of water resources risk. *Shuili Xuebao*, 2019, 50(8): 980–989.

Wang J, Church JA, Zhang X, & Chen X. Reconciling global mean and regional sea level change in projections and observations. *Nature Communications*, 2021, 12(1): 990.

Wang J, He G, Shi L, & Zhu Y. Judging the evolution of China's water use__ reading and debating the study of "Making a good judgement on the water resources demand in the new era. *China Water Resources*, 2023: 1–5.

Wang K. Application of fuzzy decision-making theory in bank-failure study. *Journal of Yangtze River Scientific Research Institute*, 2005, 22(4): 12–15.

Wang S, Cheng H, Zheng S, & Xu Q. Fluvial processes and micro-topography characteristics in the Yangtze River and Poyang Lake confluence in recent years. *Journal of Sediment Research*, 2018, 43(3): 15–20.

Wang S, Cheng H, Zheng S, Xu W, Chen G, & Zhao J. Strong erosion mechanism and trend analysis of the south Zhangjiazhou waterway of the Yangtze River. *Resources and Environment in the Yangtze Basin*, 2018, 27(9): 2070–2077.

Wang W, Fan F, Li C, & Yan J. Activity of submarine sand waves and seafloor erosion and deposition in the sea area to the southwest of Hainan Island. *Marine Geology Quaternary Geology*, 2007, 27(4): 23–28.

Wang X, Tang L, & Ding X. Analysis on the river evolution of Zhenyang reach in the Yangtze River based on GIS. *Journal of Sediment Research*, 2008, (6): 68–73.

Wang X, Dong Z, & Su˘snik J. System dynamics modelling to simulate regional water-energy-food nexus combined with the society-economy-environment system in Hunan Province, China. *Sci. Total Environ.*, 2023b, 863, 160993 https//dx.doi.org/10.1016/j.scitotenv.2022.160993.

Wang X, Xiao X, Zou Z, Hou L, Qin Y, Dong J, Doughty RB, Chen B, Zhang X, Chen Y, Ma J, Zhao B, & Li B. Mapping coastal wetlands of China using time series Landsat images in 2018 and Google Earth Engine. *ISPRS Journal of Photogrammetry and Remote Sensing*, 2020, 163, 312–326.

Wang Y, & Cui D. Time series prediction of CODMn in Dianchi Lake based on data decomposition and NARX optimization. *Pearl River*, 2024, 45(07): 92–100.

Wang Y, & Li D. Exploration of distributed law of bank collapsing and plane eddy mechanism of arc collapsing along middle-lower Yangtze River. *Rock and Soil Mechanics*, 2008, 29(4): 919–924.

Wang Y & Kuang S. Study of types and collapse modes of bank failures. *Journal of Sediment Research*, 2014, (1): 13–20.

Wang Y, & Kuang S. Study on Ωcaving mechanism in alluvial rivers. *Journal of Sediment Research*, 2006, (3): 27–34.

Wang Y, Liu Y, Jin S, Sun C, & Wei X. Evolution of the topography of tidal flats and sandbanks along the Jiangsu coast from 1973 to 2016 observed from satellites. *ISPRS Journal of Photogrammetry and Remote Sensing*, 2019, 150, 27–43.

Wang Y, Shen H, & Liu G. A review of flood channels in estuaries. *Marine Science Bulletin*, 2003, 22(3): 73–80.

Wang Y, Shen H, Li J, & Mao Z. Geomorphologic features and transport of sand waves in the flood and ebb channels of the Changjiang estuary. *Oceanologia Et Limonologia Sinica*, 2011, 42(2): 330–336.

Wang Y, Zhang Y, & Chen R. GIS-based visualized analysis of scouring and siltation in North Channel of Changjiang Estuary. *Journal of Hohai University (Natural Sciences)*, 2006, 34(2): 109–112.

Wang ZB, Maren D, Ding PX, Yang SL, Prooijen B, Vet P, Winterwerp JC, Vriend H, Stive MJF, & He Q. Human impacts on morphodynamic thresholds in estuarine systems. *Continental Shelf Research*, 2015, 111: 174–183.

Wang Z. Sediment distribution and before-dam study in middle and lower Yangtze River stability. East China Normal Univeristy, 2006.

Ward PJ, Jongman B, Aerts JCJH, Bates PD, Botzen WJW, Loaiza AD, Hallegatte S, Kind JM, Kwadijk J, Scussolini P, & Winsemius HC. A global framework for future costs and benefits of river-flood protection in urban areas. *Nature Climate Change*, 2017, 7(9): 642.

Webster PJ, Holland GJ, Curry JA, & Chang HR. Changes in tropical cyclone number, duration, and intensity in a warming environment. *Science*, 2005, 309(5742): 1844–1846.

Wieczorek M, & Migoń P. Automatic relief classification versus expert and field based landform classification for the medium-altitude mountain range, the Sudetes, SW Poland. *Geomorphology*, 2014, 206(2): 133–146.

Winterwerp JC, Wang ZB, Van BA, Van HG, & Kosters F. Man-induced regime shifts in small estuaries—II: A comparison of rivers. *Ocean Dynamics*, 2013, 63(11): 1293–1306.

Winterwerp JC, & Wang ZB. Man-induced regime shifts in small estuaries—I: Theory. *Ocean Dynamics*, 2013, 63(11): 1279–1292.

Wood J. The geomorphological characterization of digital elevation models. Univ. of Leicester, 1996.

Wu H, Zhu J, Shen J, & Wang H. Tidal modulation on the Changjiang River plume in summer. *Journal of Geophysical Research: Oceans*, 2011, 116(C8): 17.

Wu J, Wang Y, & Cheng H. Bedforms and bed material transport pathways in the Changjiang(Yangtze) Estuary. *Geomorphology*, 2009, 104(3): 175–184.

Wu L, Wang C, & Wang B. Westward shift of western North Pacific tropical cyclogenesis. *Geophysical Research Letters*, 2015, 42: 1537–42.

Wushuaihu. Morphological change of estuarine channel in response to human activities. East China Normal University, 2017.

Wu S, Cheng H, Li J & Mo C. Recent processes of morphology and micro-topography in South Passage of the Yangtze Estuary. *Journal of Sediment Research*, 2016, (5): 47–53.

Wu S, Cheng H, Li J & Zheng S. Recent morphological variation and micro-topography features in North Channel of the Yangtze Estuary. *Journal of Sediment Research*, 2014, (2): 26–32.

Wu S, Cheng H, Xu Y, Li J, Zheng S, Xu W & Lu X. Observation and analysis of bedform morphology of the Yangtze River estuary. *The Ocean Engineering*, 2016, 34(6): 65–73.

Wu SH, Cheng HQ, & Xu Y. Decadal changes in bathymetry of the Yangtze River Estuary: Human impacts and potential saltwater intrusion. *Estuarine, Coastal and Shelf Science*, 2016, 182: 158–169.

Wu SH, Cheng HQ, Xu YJ, & Li JF. Riverbed micromorphology of the Yangtze River Estuary, China. *Water*, 2016, 8(5): 1–13.

Wu Y, Yan H, & Dai W. Recent evolution and regulation of distributary channel of Beigang in Yangtze River estuary. *Yangtze River*, 2016, 47(24): 12–16+31.

Wu J, Wang Z. A hybrid model for water quality prediction based on an artificial neural network, wavelet transform, and long short-term memory. *Water*, 2022, 14(4), 610.

Wu J, Chen X, Li R, Wang A, Huang S, Li Q, & Wang Z. A novel framework for high resolution air quality index prediction with interpretable artificial intelligence and uncertainties estimation. *Journal of Environmental Management*, 2024, 357, 120785.

Wu J, Dong J, Wang Z, Hu Y, & Dou W. A novel hybrid model based on deep learning and error correction for crude oil futures prices forecast. *Resources Policy*, 2023, 83, 103602.

Wu J, Hu Y, Wu D, & Yang Z. An aquatic product price forecast model using VMD-IBES-LSTM hybrid approach. *Agriculture*, 2022, 12(8), 1185.

Wu J, Wang Z, & Dong L. Prediction and analysis of water resources demand in Taiyuan City based on principal component analysis and BP neural network. *AQUA—Water Infrastructure, Ecosystems and Society*, 2021, 70(8), 1272–1286.

Wu J, Wang Z, Dong J, Cui X, Tao S, & Chen X. Robust runoff prediction with explainable artificial intelligence and meteorological variables from deep learning ensemble model. *Water Resources Research*, 2023, 59(9), e2023WR035676.

Wu J, Wang Z, Dong J, Yao Z, Chen X, & Fan H. Multi-step ahead dissolved oxygen concentration prediction based on knowledge guided ensemble learning and explainable artificial intelligence. *Journal of Hydrology*, 2024, 131297.

Wu J, Wang Z, Hu Y, Tao S, & Dong J. Runoff forecasting using convolutional neural networks and optimized bi-directional long short-term memory. *Water Resources Management*, 2023, 37(2), 937–953.

Wu YJ, Wu SY, Wen JH, Xu M, & Tan JG. Changing characteristics of precipitation in China during 1960–2012. *International Journal of Climatology*, 2016, 36: 1387–1402.

Wulder MA, White JC, Bater CW, Coops NC, Hopkinson C, & Chen G. Lidar plots_a new large-area data collection option: Context, concepts, and case study. *Canadian Journal of Remote Sensing*, 2012, 38, 600–618.

Xi ZC, Xu H, Xing Y, Gong W, Chen G, & Yang S. Forest canopy height mapping by synergizing ICESat-2, Sentinel-1, Sentinel-2 and topographic information based on machine learning methods. *Remote Sensing*, 2022, 14, 364.

Xia J, Chen J, & She D. Impacts and countermeasures of extreme drought in the Yangtze river basin in 2022. *Shuili Xuebao*, 2022, 53(10): 1143–1153.

Xia J, Zong Q, Xu Q, & Deng C. Soil properties and erosion mechanisms of composite riverbanks in Lower Jingjiang Reach. *Advances in Water Science*, 2013, 24(6): 810–820.

Xie T, Kong R, Nurunnabi AA, Bai S, & Zhang X. Machine-learning-method-based inversion of shallow bathymetric maps using ICESat-2 ATL03 data. *IEEE Journal of Selected Topics in Applied Earth Observations and Remote Sensing*, 2023, 16, 3697–3714.

Xie Y, Sun W, Ren M, Chen S, Huang Z, & Pan X. Stacking ensemble learning models for daily runoff prediction using 1D and 2D CNNs. *Expert Systems with Applications*, 2023, 217, 119469.

Xing Y, Huang J, Gruen A, & Qin L. Assessing the performance of ICESat-2/ATLAS multi-channel photon data for estimating ground topography in forested terrain. *Remote Sensing*, 2020, 12, 2084.

Xu F, & Deng J. Qnantitative analysis on the influential factors of bank failure. *Port & Waterway Engineering*, 2005, (2): 54–57.

Xu H, Jia A, Song X, & Bai Y. Extraction and spatiotemporal evolution analysis of tidal flats in the Bohai Rim during 1984–2019 based on remote sensing. *Journal of Geographical Sciences*, 2023, 33, 76–98.

Xu H, Fan L, & Gu M. On tidal mark and tidal current mark in the Yangtze River. *Port & Waterway Engineering*, 2012, (6): 15–20.

Xu N, Ma Y, Yang J, Wang XH, Wang Y, & Xu R. Deriving tidal flat topography using icesat-2 laser altimetry and sentinel-2 imagery. *Geophysical Research Letters*, 2022, 49, e2021GL096813.

Xu W. Analysis of river regime evolution and dynamic sediment for the north inlet of North Channel in the Changjiang estuary. East China Normal University, 2016.

Xu W, Cheng H, Zheng S, Liu G, Lu X, Wu S, & Guo X. Analysis of river regime evolution and trend prediction for the north inlet of the North Channel in the Yangtze Estuary. *Marine Science Bulletin*, 2017, 36(2): 43–50.

Xu W, Cheng H, Zheng S *et al.* Evolution of Nanjing Channel in the Yangtze River and its response to human activities during the last 20 years. *Scientia Geographica Sinica*, 2019, 39(4): 663–670.

Xu Y, Liang Z, Wang X, Li W, & Du Y. Analysis on bank failure and river channel changes. *Journal of Sediment Research*, 2001, (4): 41–46.

Xu Q, Tang G, & Zhang X. Research and application of prediction model for river bank-failure. *Engineering Journal of Wuhan University*, 2004, 37(6): 9–12.

Xue H. Process and regulation of bifurcation of Southern and Northern Waterway in Yangtze River estuary. *The Ocean Engineering*, 2006, 24(1): 31–37.

Xu Z, Cheng L, Liu P, Hou Q, Cheng S, Qin S, Liu L, & Xia J. Investigating the spatial variability of water security risk and its driving mechanisms in China using machine learning. *Journal of Cleaner Production*, 2022, 362: 132303.

Yan G, Liang S, Ke Y, Liu S, Zhao H, & Yang Z. Optimizing classification of landform element using feature space: A case study in a loess area. *The Electronic Journal of Geotechnical Engineering*, 2016, 21: 5814–5824.

Yan J, Zhao S, Su F, Du J, Feng P, & Zhang S. Tidal flat extraction and change analysis based on the RF-W Model: A case study of Jiaozhou Bay, East China. *Remote Sensing*, 2021, 13(8), 1436–1436.

Yang L, Zhu J, & Zhu S. 3D numerical simulation of tide and tidal current fields in the Changjiang Estuary, Hangzhou Bay and their adjacent sea. *Journal of East China Normal University (Natural Science)*, 2001, (3): 74–84.

Yang P, Zhang S, Xia J, Chen Y, Zhang Y, Cai W, Wang W, Wang H, Luo X, & Chen X. Risk assessment of water resource shortages in the Aksu River basin of northwest China under climate change. *Journal of Environmental Management*, 2022, 305: 114394.

Yang S, Li M, & Zhang W. Sensibility of delta forward coastal response to decrease in riverine sediment supply: A case study from the seaward coast of the Chongming Island in the Changjiang River delta. *Geography and Geo-Information Science*, 2006, 22(6): 62–65.

Yang SL, Milliman JD, Li P, & Xu K. 50000 dams later: Erosion of the Yangtze River and its delta. *Global and Planetary Change*, 2011, 75: 14–20.

Yang SL, Xu KH, Milliman JD, Yang HF, & Wu CS. Decline of Yangtze River water and sediment discharge: Impact from natural and anthropogenic changes. *Scientific Reports*, 2015, 5: 12581.

Yang W, Dai R, Zhang Y, Li M, & Gong W. Analysis of variations and causes in runoff and sediment of Yangtze River during 2000–2015. *Science of Soil and Water Conservation*, 2019, 17(1): 20–27.

Yang X, & Chen L. Using multi-temporal remote sensor imagery to detect earthquake-triggered landslides. *International Journal of Applied Earth Observation and Geoinformation*, 2010, 12(6): 487–495.

Yang Z, Cheng H, Cao Z, Guo X, & Shi X. Effect of riverbed morphology on lateral sediment distribution in estuaries. *Journal of Coastal Research*, 2018, 341(1): 202–214.

Yang Z, Wu L, & Qiao J. Prediction of sewage environment based on GM-RBF. *Control Engineering of China*, 2019, 26(09): 1728–1732.

Yang ZY, Cheng HQ, & Li JF. Nonlinear advection, Coriolis force, and frictional influence in the South Channel of the Yangtze Estuary, China. *Science China Earth Science*, 2015, 58(3): 429–435.

Yang ZY, Swart HED, Cheng HQ, & Jiang CJ, Arnoldo Valle-Levinson. Modelling lateral entrapment of suspended sediment in estuaries: The role of spatial lags in settling and M4 tidal flow. *Continental Shelf Research*, 2014, 85(5): 126–142.

Yao F, Livneh B, Rajagopalan B, Wang J, Crétaux J, Wada Y, & Berge-Nguyen M. Satellites reveal widespread decline in global lake water storage. *Science (American Association for the Advancement of Science)*, 2023, 380(6646): 743–749.

Yao S, Yu W, & Dong Y. Characteristics of water flow and sediment motion in braided river and their effects on fluvial evolution. *Journal of Yangtze River Scientific Research Institute*, 2003, 20(1): 7–9.

Yao Z, Wang Z, Wang D, Wu J, & Chen L. An ensemble CNN-LSTM and GRU adaptive weighting model based improved sparrow search algorithm for predicting runoff using historical meteorological and runoff data as input. *Journal of Hydrology*, 2023, 625, 129977.

Yap L, Kande LH, Nouayou R, Kamguia J, Ngouh NA, & Makuate MB. Vertical accuracy evaluation of freely available latest high-resolution (30 m)global digital elevation models over cameroon(central africa)with gps/leveling ground control points. *International Journal of Digital Earth*, 2019, (005), 012.

Ye Z. Research on vascular plants resource and ecological protection of gucheng Lake with Xuhu wetland. Nanjing Normal University, 2014.

Yin H, Zhang X, Wang F, Zhang Y, Xia R, & Jin J. Rainfall-runoff modeling using LSTM-based multi-state-vector sequence-to-sequence model. *Journal of Hydrology*, 2021, 598, 126378.

Yin B, Xu Y, Ren L, Yang D, & Cheng M. Risk assessment of overtopping dam under waves and surges in coastal areas of the Huanghe (Yellow) river delta. *Oceanologia et Lmnologia Sinica*, 2006, 37(5): 457–463.

Yokoyama R, Shirasawa M, & Pike RJ. Visualizing topography by openness: A new application of image processing to digital elevation models. *Photogrammetric Engineering and Remote Sensing*, 2002, 68(3): 257–266.

Yu D, Tian C, & Yan Y. An analysis of current characteristics at the Yangtze River estuary. *Port & Waterway Engineering*, 2004 (1): 49–53.

Yu F. Study on flow characteristics at the confluence of the Beiyun river and the Liangshui river and the general layout of the Yulinzhuang sluice. Tsinghua University, 2016.

Yu Y, Guo M, Liu L, & Yan M. Mechanism of the change of tidal amphidromic system due to mean sea level (MSL) rise (I): Study based on the theoretical model. *Periodical of Ocean University of China*, 2008, 7(4): 517–526.

Yuan J, Xu Q, Dong B, Wang T, & Yang C. Sediment deposition of Three Gorges Reservoir and its impact on the reservoir area in recent 20 years. *Journal of Lake Science*, 2023, 35(2): 632–641.

Yuan X. Variation character of tidal dynamics along Yangtze River Estuary for the past four decades. East China Normal University, 2018.

Yuan X, Cheng H, Zheng S, Shi S, & Ma Y. Astudy on the trend and change mechanism of tidal dynamic at Datong-Nanjing Reach of the Yangtze River in recent years. *Marine Science Bulletin*, 2019, (5): 553–561.

Yue Y. The changes of future runoff and drought and flood in the Yangtze River Basin under climate change. East China Normal University, 2022.

Yun C. Comprehensive control and development strategy on the basis of the changes in water and silt conditions and riverbed topographic features. *Marine Geology Letters*, 2004, 20(7): 8–14.

Zanaga D, van de Kerchove R, Daems, D, de Keersmaecker W, Brockmann C, Kirches G, Wevers J, Cartus O, Santoro M, Fritz S, Lesiv M, Herold M, Tsendbazar N, Xu P, Ramoino F, & Arino, O. ESA WorldCover 10 m 2021 v200 (Version v200) [Data set]. Zenodo, 2022. https://doi.org/10.5281/zenodo.7254221.

Zha Y, Gao J, & SNI. Use of normalized difference built-up index in automatically mapping urban areas from TM imagery. *International Journal of Remote Sensing*, 2003, 24(3), 583–594.

Zhang EF, Savenije HHG, Chen SL, & Mao XH. An analytical solution for tidal propagation in the Yangtze Estuary, China. *Hydrology and Earth System Sciences*, 2012, 16(9): 3327–3339.

Zhang EF, Gao S, Savenije HHG, Si C, & Cao S. Saline water intrusion in relation to strong winds during winter cold outbreaks: North Branch of the Yangtze Estuary. *Journal of Hydrology*, 2019, 574: 1099–1109.

Zhang EF, Savenije HHG, Wu H, Kong Y, & Zhu J. Analytical solution for salt intrusion in the Yangtze Estuary, China. *Estuarine Coastal & Shelf Science*, 2011, 91(4): 492–501.

Zhang H, Hu L, Zhang D, Zhou Y, Ma Y, Wang XH, Wang Y, Xu M, & Xu N. A method to derive tidal flat topography in Nantong, China using MODIS data and tidal levels. *Canadian Journal of Remote Sensing*, 2021, 47(1), 17–32.

Zhang H, Wang L, Zhao Y, Cao J, & Xu M. Application of airborne LiDAR measurements to the topographic survey of the tidal flats of the Northern Jiangsu radial sand ridges in the Southern Yellow Sea. *Frontiers in Marine Science*, 2022, 9, 871156.

Zhang J, Chen X, Khan A, Zhang Y K, Kuang X, Liang X, & Nuttall J. Daily runoff forecasting by deep recursive neural network. *Journal of Hydrology*, 2021, 596, 126067.

Zhang J, Cheng H, Chen G, & Li J. Stability analysis on recent floodplain side slope in the lower reach of the Yangtze River. *Journal of Sediment Research*, 2019, 44(3): 39–46.

Zhang J, Zhou F, Cheng H, Shi S, Zhou Q, & Jiang Y. Stability analysis on arc collapsing channel slope in the lower reaches of the Yangtze River based on multimodal sensor system. *Journal of Natural Disasters*, 2018a, 27(1): 155–162.

Zhang J, Zhou F, Cheng H, Zheng S, Shi S, Jiang Y, & Zhou Q. Application of multimodal sensor system in channel slope. *Topographic Surveying*, 2018b, (3): 102–107.

Zhang JY, & Xiang Y. Analysis on the impact of climate change on the water conservancy project safety (in Chinese). *Science Sinica Technology*, 2018, 48: 1031–1039.

Zhang K, Dong X, Liu Z, Gao W, Hu Z, & Wu G. Mapping tidal flats with Landsat 8 images and Google Earth Engine: A case study of the China's Eastern Coastal Zone circa 2015. *Remote Sensing*, 2019, 11(8), 924.

Zhang L, Hu Y, Jin J, Wu C, Zhou Y, & Cui Y. Dynamic prediction of water resources carrying capacity of Chaohu Basin and system optimization regulation based on system dynamics simulation. *Journal of Lake Science*, 2021, 33(1): 242–254.

Zhang Q, Shen Z, Pokhrel Y, Farinotti D, Singh V P, Xu C, Wu W, & Wang G. Oceanic climate changes threaten the sustainability of Asia's water tower. *Nature*, 2023, 615(7950): 87–93.

Zhang S. Temporal and spatial variation of tidal range in the Changjiang Estuary and its environmental significance. East China Normal University, 2019.

Zhang W, Qi J, Wan P, Wang H, Xie D, Wang X, & Yan G. An easy-to-use airborne LiDAR data filtering method based on cloth simulation. *Remote Sensing*, 2016, 8(6), 501.

Zhang W, Zhang J, Lin R, & Zong H. Tidal response of sea level rise in marginal seas near China. *Advances in Water Science*, 2013, 24(2): 243–250.

Zhang X, Tian Y, Dong N, Wu H, & Li S. The projected futures of water resources vulnerability under climate and socioeconomic change in the Yangtze River Basin, China. *Ecological Indicators*, 2023, 147: 109933.

Zhang X, Xiao X, Wang X, Xu X, Qiu S, Pan L, Ma J, Ju R, Wu J & Li B. Continual expansion of Spartina alterniflora in the temperate and subtropical coastal zones of China during 1985–2020. *International Journal of Applied Earth Observation and Geoinformation*, 2023, 117, 103192–103192.

Zhang X, Li J, Zhu W, & Cheng H. Study on recent bed changes of erosion and siltation in Yangze River Estuary. *Haiyang Xuebao*, 2015, 37(3): 134–143.

Zhang X, Jiang C, Chen C, & Ying Q. Types and features of riverbank collapse. *Advances in Science and Technology of Water Resources*, 2008, 28(5): 66–70.

Zhang X, Jiang C, Chen C, & Ying Q. Influencing factors for bank collapse in fluvial rivers. *Journal of Hohai University (Natural Sciences)*, 2009a, 37(1): 36–40.

Zhang X, Ying Q, Chen C, & Zhang S. Generalized model study on mechanism of riverbank failure. *Shuili Xuebao*, 2009b, 40(3): 263–267.

Zhang YZ, Yao LZ, & Luo HJ. Prediction of effluent COD based on quantum weighted minimal gated unit network. *Journal of Environmental Engineering Technology*, 2023, 13(5): 1857–1864.

Zhao B. Bedforms on northern Hangzhou Bay subaqueous slope: Its implication to sea floor erosion. *Shanghai Land & Resources*, 2011, 32(3): 27–34.

Zhao Y, Wu B, Shu S, Yang L, Wu J, & Yu B. Evaluation of ICESat-2 ATL03/08 surface heights in urban environments using airborne LiDAR point cloud data. *IEEE Geoscience and Remote Sensing Letters*, 2022, 19, 1–5.

Zheng SW, Cheng HQ, Shi S, Xu W, Zhou Q, Jiang Y, Zhou F, & Cao M. Impact of anthropogenic drivers on subaqueous topographical change in the Datong to Xuliujing reach of the Yangtze River. *Science China Earth Sciences*, 2018b, 61(7): 940–950.

Zheng SW, Cheng HQ, Shi SY, Xu W, Zhou QP, Jiang YH, Zhou FN, & Cao MX. Impact of anthropogenic drivers on subaqueous topographical change in the Datong to Xuliujing reach of the Yangtze River. *Science China Earth Sciences*, 2018, 61(7): 940–950.

Zheng SW, Cheng HQ, Wu S, Liu G, Lu X, & Xu W. Discovery and implications of cate-nary-bead subaqueous dunes. *Science China Earth Sciences*, 2016, 59: 495–502.

Zheng SW, Cheng HQ, Wu S, Shi S, Xu W, Zhou Q, & Jiang Y. Morphology and mechanism of the very large dunes in the tidal reach of the Yangtze River, China. *Continental Shelf Research*, 2017, 139: 54–61.

Zheng SW, Xu YJ, Cheng HQ, Wang B, & Lu X. Assessment of bridge scour in the lower, middle, and upper Yangtze River estuary with riverbed sonar profiling techniques. *Environmental Monitoring & Assessment*, 2018a, 190(1): 15.

Zheng SW, Xu YJ, Cheng HQ, Wang B, Xu WJ, & Wu SH. Riverbed erosion of the final 565 kilometers of the Yangtze River (Changjiang) following construction of the Three Gorges Dam. *Scientific Reports*, 2018, 8(1): 11917.

Zheng W, & Zhao D. Effect analysis on shoal protection and discharge control works of South and North Channel bifurcation in Yangtze Estuary. *China Harbour Engineering*, 2010, (5): 10–14.

Zhou F. On shallow spot control standard in edge area of deepwater channel in the Yangtze estuary. *Port & Waterway Engineering*, 2012, (8): 129–134.

Zhou J, Li P, Shi L, & Liu L. Impact of sea level variation on determination of wharf deck elevation. *Port & Waterway Engineering*, 2019(5): 60–66+90.

Zhou Y. Real-time probabilistic forecasting of river water quality under data missing situation: Deep learning plus post-processing techniques. *Journal of Hydrology*, 2020, 589, 125164.

Zhou Y, Cheng H, Ta N, Jiang H, Ruan R, & Zhao M. Research on early warning system for fresh water supply in Shanghai on rising sea level background. *Resources Science*, 2012, 34(7): 1312–1317.

Zhou Y, Li Qi, & Gu S. Return period of low tide level at Wusongkou Station in the lower reaches of Huangpu River. *Water Resources and Hydropower Engineering*, 2019, 50(9): 69–74.

Zhou Y, Lu N, Hu H, & Fu B. Water resource security assessment and prediction in a changing natural and social environment: Case study of the Yanhe Watershed, China. *Ecological Indicators*, 2023, 154: 110594.

Zhou Y, & Pan J. Experimental studies on stability of perforated concrete block revetment under wave action. *Hydro-Science and Engineering*, 2016(3): 126–131.

Zhou Y, Pan J, & Zuo Q. Experimental study on the overtopping of the artificial island of Hong Kong-Zhuhai-Macao Bridge. *Advances in Water Science*, 2019, 30 (6): 908–914.

Zhu D, & Chang Y. Urban water security assessment in the context of sustainability and urban water management transitions: An empirical study in Shanghai. *Journal of Cleaner Production*, 2020, 275: 122968.

Zhu J, Cheng X, Li L, Wu H, & Lyu H. Dynamic mechanism of an extremely severe saltwater intrusion in the Changjiang estuary in February 2014. *Hydrology and Earth System Sciences*, 2020, 24(10): 5043–5056.

Zhu L, Li M, & Chen Y. Prediction model for effluent COD in sewage treatment based on BP neural network optimized by EHO. *Journal of Chongqing Technology and Business University (Natural Science Edition)*, 2022, 39(3): 26–32.

Zhong L, Xu G, & Zeng F. Immovable bed experiment study on the resistance of sand wave characterized by fractal dimension. *Journal of Basic Science and Engineering*, 2013, 21(1): 116–126.

Zong Q, Xia J, Deng C, & Xu Q. Modeling of the composite bank failure process using BSTEM. *Journal of Sichuan University (Engineering Science Edition)*, 2013, 45(3): 69–78.

Zuo J, Zuo C, Li J, & Chen M. Advances in research on sea level variations in China from 2006 to 2015. *Journal of Hohai University (Natural Sciences)*, 2015, 43(5): 442–449.

Zuo L, & Lu Y. Effect of nodes on branch evolution in Madang reach in the downstream of Yangtze River. *Journal of Yangtze River Scientific Research Institute*, 2014, 31(10): 72–79.

Zscheischler J, Westra S, van den Hurk BJJM, Seneviratne SI, Ward PJ, Pitman A, Aghakouchak A, Bresch DN, Leonard M, Wahl T, & Zhang X. Future climate risk from compound events. *Nature Climate Change*, 2018, 8(8): 469.

Index

W

X

Y

East China Normal University Scientific Reports

Published (continued from page ii)

Subseries of Symposia and Topic Studies

Published

Vol. 16 *Forty Years of Algebraic Groups, Algebraic Geometry, and Representation Theory in China: In Memory of the Centenary Year of Xihua Cao's Birth*
edited by Jie Du (University of New South Wales, Australia), Jianpan Wang (East China Normal University, China) and Lei Lin (East China Normal University, China)

Subseries on Educational Information Technology

Published

Vol. 15 *Achieving Greater Educational Impact through Data Intelligence: Practice, Challenges and Expectations of Education*
by Bian Wu (East China Normal University, China), Yiling Hu (East China Normal University, China) and Xiaoqing Gu (East China Normal University, China)

Vol. 14 *Design and Development of a Wiki-based Collaborative Process Writing Pedagogy: Putting Technological, Pedagogical, and Content Knowledge (TPACK) in Action*
by Xuanxi Li (Huzhou University, China)

Subseries on Data Science and Engineering

Published

Vol. 13 *Load Balance for Distributed Real-time Computing Systems*
by Junhua Fang (Soochow University, China), Rong Zhang (East China Normal University, China) and Aoying Zhou (East China Normal University, China)

Vol. 12 *Biological Language Model: Theory and Application*
by Qiwen Dong (East China Normal University, China), Xiaoyang Jing (Yunnan University, China), Aoying Zhou (East China Normal University, China) and Xiuzhen Hu (Inner Mongolia University of Technology, China)

Vol. 11 *Probabilistic Approaches for Social Media Analysis: Data, Community and Influence*
by Kun Yue (Yunnan University, China), Jin Li (Yunnan University, China), Hao Wu (Yunnan University, China), Weiyi Liu (Yunnan University, China) and Zidu Yin (Yunnan University, China)

Vol. 10 *Clustering and Outlier Detection for Trajectory Stream Data*
by Jiali Mao (East China Normal University, China), Cheqing Jin (East China Normal University, China) and Aoying Zhou (East China Normal University, China)

Vol. 9 *Concurrency Control and Recovery in OLTP Systems: High Scalability and Availability*
by Peng Cai (East China Normal University, China), Jinwei Guo (East China Normal University, China) and Aoying Zhou (East China Normal University, China)

Vol. 8 *Network Data Mining and Analysis*
by Ming Gao (East China Normal University, China), Ee-Peng Lim (Singapore Management University, Singapore) and David Lo (Singapore Management University, Singapore)

Vol. 7 *Time-Aware Conversion Prediction for E-Commerce*
by Wendi Ji (East China Normal University, China), Xiaoling Wang (East China Normal University, China) and Aoying Zhou (East China Normal University, China)

Vol. 6 *Discovery and Fusion of Uncertain Knowledge in Data*
by Kun Yue (Yunnan University, China), Weiyi Liu (Yunnan University, China), Hao Wu (Yunnan University, China), Dapeng Tao (Yunnan University, China) and Ming Gao (East China Normal University, China)

Vol. 5 *Review Comment Analysis for E-commerce*
by Rong Zhang (East China Normal University, China), Aoying Zhou (East China Normal University, China), Wenzhe Yu (East China Normal University, China), Yifan Gao (East China Normal University, China) and Pingfu Chao (East China Normal University, China)

Vol. 4 *Opinion Analysis for Online Reviews*
by Yuming Lin (Guilin University of Electronic Technology, China), Xiaoling Wang (East China Normal University, China) and Aoying Zhou (East China Normal University, China)

Vol. 3 *Querying and Mining Uncertain Data Streams*
by Cheqing Jin (East China Normal University) and Aoying Zhou (East China Normal University

Directly Under the Series

Vol. 2 *School Mathematics Textbooks in China: Comparative Study and Beyond*
by Jianpan Wang (East China Normal University)

Vol. 1 *High-Frequency Trading and Probability Theory*
by Zhaodong Wang (East China Normal University) and Weian Zheng (East China Normal University)

East China Normal University Scientific Reports

www.ingramcontent.com/pod-product-compliance
Lightning Source LLC
LaVergne TN
LVHW011809200926
848484LV00005B/1120